AIR POLLUTION

PART B—PREVENTION AND CONTROL

ǀ

ENVIRONMENTAL HEALTH ENGINEERING TEXTBOOKS

CONSULTING EDITORS:
EARNEST F. GLOYNA and *JOE O. LEDBETTER*

Volume 1:   PRINCIPLES OF RADIOLOGICAL HEALTH
 by EARNEST F. GLOYNA and JOE O. LEDBETTER
Volume 2:   AIR POLLUTION *(in two parts)*
 by JOE O. LEDBETTER

OTHER TEXTBOOKS IN PREPARATION

AIR POLLUTION

IN TWO PARTS

Part B—Prevention and Control

JOE O. LEDBETTER

COLLEGE OF ENGINEERING
THE UNIVERSITY OF TEXAS
AUSTIN, TEXAS

MARCEL DEKKER, INC. New York   1974

MARCEL DEKKER, INC.
270 Madison Avenue, New York, New York 10016

LIBRARY OF CONGRESS CATALOG CARD
NUMBER: 71-157837

ISBN: 0-8247-1406-7

Current printing (last digit):
10 9 8 7 6 5 4 3 2 1

Printed in the United States of America

To my brothers and sisters, by blood and by law; the former for helping to rear me and all of them for being so good to me for many years.

Del Whitaker

Gladys Hammett

Oscar and Brenda Ledbetter

Smith and Elizabeth Ledbetter

Sue Thompson

Wendell and Vernie Thompson

W. L. and Lucille Thompson

PREFACE

This book is the second part of a two-part set on the subject of air pollution. Both volumes are intended to serve a need, as I see it, in the literature that treats the engineering control of air pollution.

The first of the two volumes, Air Pollution, Part A: Analysis, contains ten chapters on those topics essential for analysis of air pollution, analysis in the sense of running tests and evaluating the consequences of given pollution levels. This second volume uses fourteen chapters, numbered eleven through twenty-four, plus appendixes to cover the material deemed necessary for the effective engineering of air pollution control. The chapters are grouped into five functional sections--Planning, Prevention, Gaseous Control, Particulate Control, and Specific Controls.

Section I, Planning, is composed of three chapters. Chapter 11 briefly describes the subject of air pollution management along the lines being followed in the United States. It reflects the belief of the author that esthetics, vegetation damage, endangerment of wildlife, and materials effects, not human health effects, are the real reasons for air pollution control. Tenuous correlations, probably accidental, between air pollution and dis-ases have been taken as proof of cause-effect, a practice that led the eminent statistician William Feller to write: "The others [the many negative correlations] are mercifully forgotten, but the scandal is that the 'significant' results are published as though they had meaning. ...it is saddening that also this black magic passes for art."

That information required to make control decisions is discussed in Chapter 12 under the topical areas of gas stream characteristics, estimation of costs, and selection of abatement method.

Chapter 13 deals with ancillary operations that are essential aspects of control methods. The pollutant must be collected at or near its source by an exhaust system; the exhausted gas stream may require conditioning, such as cooling; the polluting material, once removed, has to be handled and disposed of; and peripheral equipment is necessary for the air cleaning process.

Section II describes the subject of Prevention in two chapters, one on preventing the formation and/or release of the pollutant and the other on dispersing the pollutant to prevent problem ground-level concentrations. The latter method has been particularly well suited for use with sulfur dioxide from power plants and other dilute sources for which removal was impractical; despite the large number of processes that have been developed for the removal of dilute sulfur dioxide, such removal is still infeasible.

Section III has three chapters on Gaseous Control. Absorption and adsorption are presented in Chapters 16 and 17 with a format that is typical for the chapters on air cleaning. Each chapter has paragraphs of introduction, theory, design, applications, costs, and summary. In addition, a set of related problems and a selected bibliography are included.

Chapter 18 discusses combustion in the format listed plus the inclusion of a topic on condensation and other methods for the control of gaseous pollutants.

Five chapters on Particulate Control comprise Section IV. The three types of air-cleaning processes that are capable of generally meeting today's rigid emission regulations are covered in chapters on filtration, scrubbing (high energy), and electrostatic precipitation. Inertial cleaning has very limited application in modern control because of its inability to collect small particles efficiently, but it does retain significance in air cleaning and is discussed in Chapter 19. Chapter 23 gives a brief description of the techniques and methods which find less use.

Section V has a single chapter on the subject of Specific Controls for some of the common air pollution types and sources. Chapter 24 should point the way to application of the information in other chapters to a given problem.

The proper degree of air pollution control is clearly specified in current Federal law which says "the best available technology that is feasible" must be used. The zealots who would have us ignore feasibility in the name of unproved and unprovable health effects do not, in the long-run, serve the real cause of air pollution control any more than those who ignore the "appropriate tests" part of the Delaney clause in the Food, Drugs, and Cosmetics Act in order to "prove" that any food additive causes cancer. We need to guard diligently against letting the "what if" group decide our control strategies by their emotions. After all, the secondary standards which are set for the purposes described here as the real reasons for control are as strict as, or stricter than those levels prescribed by the primary standards to protect the health. Let the clearer heads prevail in following the admonition of James Thurber, "A pinch of probably is worth a pound of perhaps."

Here again, I wish to acknowledge the contributions to this work by my wife, students, and colleagues.

Austin, Texas   Joe O. Ledbetter
January 1974

CONTENTS

PREFACE   v

Section I: Planning   1

CHAPTER 11. MANAGEMENT OF AIR POLLUTION   3
I.   Introduction   3
II.   Criteria   4
III. Goals   9
IV.   Standards   12
V.   Emission Limits   16
VI.   Summary   18
      Problems   19
      Bibliography   20

CHAPTER 12. BASES FOR AIR POLLUTION ABATEMENT
                                        DECISIONS   23
I.   Introduction   23
II.   Characteristics of the Gas Streams   23
III. Estimating Costs   29
IV.   Selection of Abatement Method   35
V.   Summary   45
      Problems   46
      Bibliography   47

CHAPTER 13. AUXILIARY OPERATIONS   49
I.   Introduction   49
II.   Local Exhaust Systems   49
III. Conditioning Gas Stream   60

IV.   Handling and Disposal of Collected Material   65
V.   Equipment for Peripheral Operations   66
VI.   Summary   68
       Problems   69
       Bibliography   69

Section II: Prevention   71

CHAPTER 14. PREVENTING FORMATION/RELEASE
                                    OF POLLUTANT   73
I.   Introduction   73
II.   Process Substitution   73
III.   Enclosing Process   76
IV.   Relocation of Process   77
V.   Summary   77
       Problems   78
       Bibliography   78

CHAPTER 15. TALL STACKS--DISPERSION FOR DILUTION
                                    AND DECAY   79
I.   Introduction   79
II.   Theory   79
III.   Design   81
IV.   Applications   92
V.   Costs   93
VI.   Summary   94
       Problems   94
       Bibliography   95

Section III: Gaseous Control   97

CHAPTER 16. ABSORPTION   99
I.   Introduction   99
II.   Theory   99

CONTENTS

III.   Design   105
IV.   Applications   117
V.   Costs   119
VI.   Summary   119
       Problems   120
       Bibliography   120

CHAPTER 17. ADSORPTION   123
I.   Introduction   123
II.   Theory of Adsorption   123
III.   Design of Adsorbers   124
IV.   Applications   130
V.   Costs   131
VI.   Summary   132
       Problems   132
       Bibliography   133

CHAPTER 18. COMBUSTION AND OTHER METHODS   135
I.   Introduction   135
II.   Combustion   135
III.   Other Methods   147
IV.   Summary   148
       Problems   148
       Bibliography   149

Section IV: Particulate Control   151

CHAPTER 19. GRAVITATIONAL, CENTRIFUGAL, AND
                                INERTIAL COLLECTION   155
I.   Introduction   155
II.   Gravitational Collection   155
III.   Centrifugal Collection   159
IV.   Inertial Collection   164
V.   Summary   165
       Problems   166
       Bibliography   166

CHAPTER 20. FILTRATION   167
I.   Introduction   167
II.   Theory   167
III.   Design   169
IV.   Applications   182
V.   Costs   185
VI.   Summary   186
        Problems   187
        Bibliography   187

CHAPTER 21. SCRUBBING   189
I.   Introduction   189
II.   Theory   189
III.   Design   190
IV.   Applications   200
V.   Costs   201
VI.   Summary   201
        Problems   202
        Bibliography   202

CHAPTER 22. ELECTROSTATIC PRECIPITATION   205
I.   Introduction   205
II.   Theory   205
III.   Design   207
IV.   Applications   217
V.   Costs   218
VI.   Summary   219
        Problems   220
        Bibliography   220

CHAPTER 23. AIR-CLEANER COMBINATIONS AND
                              OTHER METHODS   223
I.   Introduction   223
II.   Air-Cleaner Combinations   223

III. Other Methods 226
IV. Summary 226
　　Problems 227
　　Bibliography 228

Section V: Specific Controls 229

CHAPTER 24. SPECIFIC POLLUTION CONTROLS 231

I. Introduction 231
II. Sulfur Dioxide 231
III. Oxides of Nitrogen 236
IV. Auto Exhaust 238
V. Metallurgical Fumes 242
VI. Acid Mists 243
VII. Noise 243
VIII. Other Pollution 244
IX. Summary 247
　　Problems 248
　　Bibliography 248

APPENDIX A. PROBLEM SOLUTIONS FOR PART A 251

APPENDIX B. PROBLEM SOLUTIONS FOR PART B 259

APPENDIX C. SOME SOURCES OF AIR POLLUTION EQUIPMENT 275

I. Sampling Equipment 275
II. Analytical Equipment 275
III. Gaseous Control 276
IV. Particulate Control 277

AUTHOR INDEX 279

SUBJECT INDEX 283

AIR POLLUTION

PART B—PREVENTION AND CONTROL

Section I: Planning

Planning for air pollution abatement is highly important to
the successful attainment of desirable air quality. Abatement
may take the form of preventing the formation and/or release of the
pollutant, dispersing the pollutant releases through a tall stack,
dispersing the pollutant sources, or cleaning the waste gas stream
before releasing it into the atmosphere. If air cleaning is employed,
the collected pollutant must be disposed of without causing additional problems. The choice of abatement method clearly lies in
the planning area. The degree of air pollution control that should
be asked for is undoubtedly the most perplexing problem in this
field today.

Chapter 11

MANAGEMENT OF AIR POLLUTION

I. INTRODUCTION

In order to bring air pollution under effective control, it is
necessary to define the degree of control needed. The task of
making this definition is exceedingly difficult. Adding to the
difficulty of setting limits is the fact that different regions need
different controls because of varying land use, topographical
features, and meteorological conditions; yet, the copying of regulations from one geographical area to another and the setting of
single regulations for whole states or all of the United States
continue to be common practice. Furthermore, some of the controls that are badly needed in areas where stagnating periods
produce photochemical smog problems are probably superfluous
in the plains states and in Texas where stagnating periods are
of short duration (see Fig. 3-6a).

The management sequence includes writing meaningful
criteria for the effects caused by various levels of ambient air
pollution, setting goals for the quality of air sought, adopting
realistic standards for the short-range steps toward the goals,
and enacting enforceable regulations on the emission limits as
required to meet the standards. The criteria, standards, and
emission limits will necessarily change as the sociological,
technological, and epidemiological changes occur; however,
goals must not be relaxed without strong cause and they should
have been set strong enough not to need stiffening for a long time.

II. CRITERIA

Criteria are the bases which are used in the setting of air quality control measures. They are written to summarize the current knowledge of the effects the air pollutants.

A. How to Write

Criteria should be based on the best scientific evidence of effects that were probably caused by air pollutants. Because of the epidemiological procedures, only statistically reliable health data should be used. Many superficial studies report meaningless conclusions on effects; such conclusions have a way of prevailing for a long time before they are refuted.

It is highly important to do a good job on the presentation of the physical and vegetative effects as well as physiological effects on man. The physical and vegetative effects are generally much more definitive at the present than the human health effects and probably provide a better basis for most ambient air standards. Criteria should not be written to justify preconceived control levels. Opinions of the highly qualified may be used in criteria if they are clearly labeled and presented with the data that have been found in practice.

B. Toxicological/Epidemiological Dilemma

The principal obstacle to meaningful criteria based on the epidemiological data for chronic ambient exposures is the apparently contradictive toxicological data from tests on the same pollutant. Toxicology principles have long been used for industrial hygiene control of occupational illness without apparent damage from exposures which are orders of magnitude higher than those for ambient air pollution. Most toxicologists believe in a threshold level of exposure, a level below which the risk of any physiological damage is negligible (see Fig. 11-1). If it is true as Paracelsus said in 1530 that "Dose alone makes poison," why

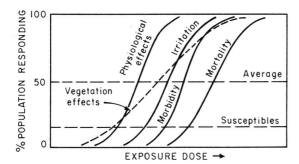

FIG. 11-1. Susceptibility variations. From "Rationale for Air Quality Criteria," Environmental Sci. Tech., 2:10, 742-748, October 1968.

do the conclusions drawn from epidemiological studies indicate disease from air pollution exposures?

The Arndt-Schulz law for the physiological effect of toxic materials accommodates the toxicological viewpoint very well; i. e., large doses destroy, moderate doses inhibit, and small doses stimulate. The stimulative dose for any given poison is a very large dose as contrasted to ambient dosage levels; it may be as high as a few or several percent of the median lethal dose (LD_{50}). Apparently the stimulatory dose causes the body to "over-repair" the damage done by the toxic material. This phenomenon has shown up in many experiments which measured the effects of poisons, ionizing radiations, and other physiological stresses on growth rate, size, length of life, and other yardsticks of what is considered as physiologically good. A recent example was some work done by scientists at a reliable laboratory on the effects of sulfur dioxide on guinea pigs. Continuous exposures for 1 year at 0.1, 1.0, and 5 ppm_v of sulfur dioxide showed no deleterious effects from the exposures. Indeed, those animals exposed at 5 ppm_v exhibited a higher survival rate than any of the other groups, including the controls, and these animals also had reduced incidence and severity of pulmonary changes found in histological examinations.

Apparently a fourth stated level would need to be added to the Arndt-Schulz law in order for the law to apply in epidemiology. This statement would be that very small exposures (doses) over a very large population for a long time will cause very small, but statistically detectable, numbers of illnesses. Such a statement ignores the frequent misuse of statistics in drawing epidemiological conclusions. Most attempted correlations of environmental stress with disease go unreported because they did not show the presupposed link. Many of the correlations that are found result merely from the fact that 5% of the attempts will show significance at the 0.05 level. Logic indicates that a large portion of the reported cause-effect relationships from correlations are not true measures of anything other than statistical fluctuations. Criteria which do not present both the toxicological and epidemiological data and the differences between the two are not likely to be widely accepted and will be subject to criticism, perhaps ridicule.

C. Some Current Criteria

There is another philosophical rift in current criteria besides the toxicology-epidemiology one just discussed; it is whether to use physiological damage or physiological change as the guide-line for adverse effects criteria. The American and European practices have been to cite the exposure levels which will cause physiological damage,whereas Russian practice has been to cite exposures which cause any detectable physiological change-- changes indicated by an increase in the blink rate, a decrease in the reading rate or other psychomotor function tests, or Pav-lovian response. The physiological change philosophy was used in the Air Quality Criteria for Carbon Monoxide issued by the National Air Pollution Control Administration.

1. Carbon Monoxide

The AQC for carbon monoxide cited above gives the follow-ing principal conclusions about the health effects of carbon monoxide exposures: 10 to 15 ppm_V for 8 hr (2 to 2.5% carboxyhemo-globin, COHb) is associated with impaired time-interval discrimination; 30 ppm_V for 8 hr (5% COHb) impairs psychomotor tests in nonsmokers; and some evidence indicates that 8 to 14 ppm_V weekly average levels cause an increase in mortality among myocardial infarction patients.

Various investigations of the toxicological aspects of carbon monoxide have shown that no lasting damage could be found from repeatedly exposing men to levels of carbon monoxide which caused up to 40% COHb. For exposures less than a few hours in duration, the effects are: 300 ppm_V-hr, none noticeable; 600 ppm_V-hr, barely noticeable; 900 ppm_V-hr, headache and nausea; and 1500 ppm_V-hr, dangerous to life. The effects of carbon monoxide are essentially those associated with anoxia and are, therefore, synergistic with the effects of high altitude and smoking. A heavy smoker may show a level of COHb as high as 10% solely from the exposure to the 200 to 800 ppm_V of carbon monoxide in the smoke.

There are no proved effects other than anoxia in animals attributable to carbon monoxide--no physical, chemical, or vegetative effects.

2. Sulfur Oxides

The AQC for Sulfur Oxides concludes that: (1) 0.04 ppm_V (115 $\mu g/m^3$) of sulfur dioxide as an annual mean, accompanied by smoke concentrations of about 160 ug/m^3, may increase the mortality from bronchitis and lung cancer; (2) 0.21 ppm_V of sulfur dioxide as a 24-hr mean, with smoke concentrations of about 300 $\mu g/m^3$, may accentuate the symptoms in patients with chronic lung disease; (3) various levels of sulfur dioxide from 0.11 to 0.52 ppm_V may cause increases in mortality or morbidity; (4) 0.10 ppm_V of sulfur dioxide, with comparable concentration of particu-

AIR POLLUTION

late matter on a $\mu g/m^3$ basis and relative humidity of 50%, may reduce visibility to about 5 miles; (5) 0.12 ppm_v of sulfur dioxide, accompanied by high particulate levels, may increase the corrosion rate for steel panels by 50%; (6) vegetative effects of sulfur dioxide exposures may be as follows: 0.03 ppm_v annual mean, chronic plant injury and excessive leaf drop; 0.3 ppm_v for 8 hr, injury to some trees; 0.05 to 0.25 ppm_v for a few hours, synergistic with ozone or nitrogen dioxide for plant damage.

The human health effects of sulfur dioxide which have shown up in toxicological investigations are that: 10 ppm_v will cause trouble only in very sensitive individuals; 5 ppm_v causes some increase in breathing resistance; 0.2 ppm_v can cause some detectable change in conditioned reflexes centered in the cerebral cortex; and the threshold to give a recognizable odor is about 0.5 ppm_v.

The level of sulfur trioxide, as well as that for sulfuric acid, is usually tied to the sulfur dioxide and the effects need not be separated for ambient criteria. The sulfur trioxide content of the air is usually 0.01 to 0.04 times that for sulfur dioxide, with the tendency toward the smaller value. The potentiation work by Amdur revealed that the particles convert the sulfur dioxide to sulfur trioxide which goes to the acid mist to cause the enhanced damage.

3. Photochemical Oxidant (and Ozone)

The AQC for Photochemical Oxidants lists the following principal effects for this class of pollutants: (1) 0.03 to 0.3 ppm_v as an hourly average impaired the performance of student athletes; (2) 0.13 ppm_v peak (0.05 to 0.06 ppm_v hourly average) increased the asthma symptoms in some but not all sufferers of this malady; (3) 0.10 ppm_v peak (0.03 to 0.05 ppm_v hourly average) caused eye irritation; and (4) vegetation can be very sensitive to photochemical oxidants.

Toxicology experiments on acute effects have shown that: 0.1 ppm_v of ozone can cause drying of the mucous membranes of the nose, mouth, and throat; general irritation occurs above about 1 ppm_v; 1.5 to 2 ppm_v for more than 1 hr cause pulmonary edema as evidenced by reduced vital capacity; higher concentrations for an hour or more may cause pulmonary congestion, edema, and hemorrhage; and the median lethal dose (LCt_{50}) for rats is about 25 ppm_v-hr (6 X 4). Chronic effects have been shown for 1-ppm_v exposures over a long period. For example, repeated daily exposures for a year led to early emphysematous and fibrotic changes (accelerated aging) in the lungs. Ozone increased the susceptibility to streptococcal infections in test animals. It is believed to result in a hardening of the alveolar sacs and a loss in their elasticity.

Damage to tobacco plants has been observed from a 2-hr exposure to 0.053 to 0.066 ppm_v of ozone, and 4 hr at the same levels caused emergence tip burn of white pine needles. A 5-hr exposure of 0.1 ppm_v ozone did moderate damage to petunias and pinto bean plants.

Ozone hastens the destruction of rubber, the fading of dyes, and the "rotting" of many fabrics. In addition, it has a low threshold odor (0.02 to 0.05 ppm_v) and has an odor which is generally called disagreeable.

III. GOALS

The goals for a given region should be those levels desired by an informed citizenry of the region. Goals may tend toward the idealistic; however, the goals set should be probably attainable over a period of several years, and there should be definite long range plans for meeting the goals. The goals may be met by one, two, or more steps in the standards (see Fig. 11-2).

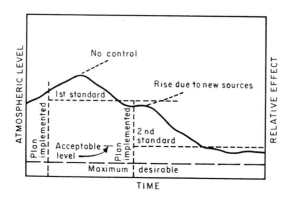

FIG. 11-2. Stepwise implementation of goals. Adapted
from "Rationale for Air Quality Criteria," Environmental Sci.
Tech., 2:10, 742-748, October 1968.

It is in formulating the goals in the ambient air quality for
a region that the people in the region should get involved. They
should be told what various goals will mean in terms of the air
quality which they can expect and they should be told in a manner
that they can understand, such as visibility, dirtying capability,
lifetime of house paints, probable plant damage, and factors with
which they have contact every day. At the same time the people
of the region should be informed as to the cost of obtaining the
various qualities of air--how many dollars and/or cents per month
will be added to their electric bills; how much new car controls
are costing and how much good they are doing; whether there are
industries which cannot economically operate at a given emission
level; and whether a certain goal is unattainable at any price short
of changing our way of life, i.e., the irreducible minimum levels
of air pollutants for the region and the possible and the practical
minimums (see Fig. 11-3).

The goals must be set with due cognizance taken of the
ambient levels that would be present as a result of uncontrollable
sources. For example, the normal activities of people in a densely
populated region may cause an average annual suspended particu-

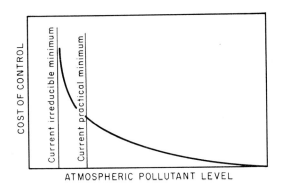

FIG. 11-3. Attainable pollutant concentrations.

lates concentration of $40 \mu g/m^3$. It would be foolhardy to set a goal of $30 \mu g/m^3$ when the implementation of complete control for all controllable sources could not meet the goal. Some of the uncontrollable particulate sources include the wear of tires on the road, the wear of the roadbed from the tires, dusts generated and those resuspended by walking, and agricultural and natural· dusts. Also, it is very expensive to remove the last vestige of a pollutant from a gas stream; the cost is much more than the benefit derived.

Idealistically, if both the value of all the benefits from air cleaning, including esthetic, recreation, and other intangibles, and the costs of each increment of the cleaning could be ascertained, a curve such as that shown in Fig. 11-3 could be used to make the decision about the level to set as a goal. In practice, considerably less than total information about the costs and benefits is available, but the decision must be made now on setting goals. Only the people directly involved can come close to weighing esthetic values to themselves. These people must help in setting goals which are meaningful. Errors made in setting the goals will probably become more evident in the future when more information is available. The goals can, with good cause, be changed. This does not suggest that economic reasons should compromise the other values set into the goals; as a result of improving technology, any goal changing in the future should be toward lower ambient levels of pollution--changes in the upward direction should receive most careful attention before they are instituted.

IV. STANDARDS

Standards are those maximum ambient levels of air pollutants which are adopted to meet, or as a step toward meeting, the goals set for the region. The standards are set by the enforcement agency which then monitors to see that they are being met.

A. Considerations for Standards

Standards should be viewed as interim in nature. They should be lowered as sociological, technological, and economical changes warrant. Standards should be attainable within a period of not more than a few years and preferably within a year. If the standards are set unreasonably low, the emissions limits either cannot be set low enough to meet the standards or cannot be enforced. Unenforceable regulations are often said to be worse than no regulations at all because they create a contempt for all regulations. The standards should be attainable at a reasonable cost with current technology. They should not anticipate a break-through in cleaning methods because of overly optimistic descriptions of a method which has not been proved in prototype.

Proposed standards should be described to those people who took part in setting the goals for the region. It should be explained to them how the standards will meet the goals and when. Short-comings of the standards should be pointed out. The goals were probably set with the desired average conditions as a basis, whereas the standards have come to mean the maximum concentrations to be encountered at any point rather than the normal;

therefore, the standard may give a larger number than the goal and still be consistent with it. Similar temporal fluctuations in the concentrations are recognized in standards which allow occasional excursions above the limits and a higher 24-hr maximum concentration than the annual mean concentration.

When setting the standards for a region, the regulating authorities must be cognizant of the levels which now prevail for the pollutant under consideration. They must know not only the ambient concentrations at various places throughout the region but also the sources of the pollutant and which sources could be eliminated or significantly reduced most easily. Only through the proper application of such knowledge can the control board promulgate realistic standards.

B. Some Current Standards

Federal air quality standards have been set for the commonly considered air pollutants--particulates, sulfur dioxide, carbon monoxide, nitrogen oxides, photochemical oxidants, and hydrocarbons (non-methane). Table 11-1 presents a brief summary of these standards. The primary standards were set for the avowed purpose of protecting human health and are to be met by 1975; the secondary standards were set to protect public welfare and are to be met within a few years after 1975.

The states may set stricter air quality standards than the Federal standards, and some of the states have done so. For example, states where natural gas is the principal fuel can generally meet the secondary standards for some of the pollutants now. Also, some standards for acid mists, fluorides, hydrogen sulfide, mercaptans, mercury, beryllium, asbestos, and other pollutants have been set by the states before any Federal action on these substances. Some states have varied standards by land use within the state. Employing such standards can prevent serious degradation of the clean areas while the dirty areas are being cleaned up to meet the maximum tolerable pollution level.

TABLE 11-1

Federal Ambient Air Quality Standards[a]

Pollutant	Time Concentration	Primary $(\mu g/m^3/ppm_v)$	Secondary
Particulates	Annual geom. mean	75	60
	Maximum 24-hr[b]	260	150
Sulfur oxides	Annual arith. mean	80/0.03	60/0.02
(sulfur dioxide)	Maximum 24-hr	365/0.14	260/0.10
	Maximum 3-hr		1300/0.50
Carbon monoxide	Maximum 8-hr	10/8.7[c]	10/8.7[c]
	Maximum 1-hr	40/35[c]	40/35[c]
Photochemical oxidants (as ozone)	Maximum 1-hr	160/0.08	160/0.08
Hydrocarbons--non- methane (as methane)	Maximum 3-hr (6-9 a.m.)	160/0.24	160/0.24
Nitrogen dioxide	Annual arith. mean	100/0.05	100/0.05

[a]"Rules and Regulations," Federal Register, 36:84, 8187, April 30, 1971.

[b]Not to be exceeded more than once per year if listed "Maximum."

[c]Carbon monoxide concentrations are mg/ppm_v.

The Environmental Protection Agency has discouraged the land use concepts by setting single standards countrywide; however, the nondegradation ruling by the Supreme Court will in effect bring about multiple standards and at least a form of land use planning.

The American Industrial Hygiene Association (AIHA) has published "Community Air Quality Guides" for some pollutants. The AIHA states its objective for the guides as "controlling air pollution so that the community is a desirable place in which to live." They recommend a 30-day particulate limit of 120 $\mu g/m^3$. If this value is extrapolated to an equivalent 24-hr basis by $C = C_0 (t/t_0)^{-0.2}$ (see Chapter 6), a limit of 240 $\mu g/m^3$ is obtained. This number compares favorably with the Federal primary standard of 260 $\mu g/m^3$; however, those of us in the cleaner areas of the country would be unwilling to accept a level of dirtiness that would, among other effects, limit the visibility to 6 or 8 km.

A logical basis for deriving air quality standards from the Threshold Limit Values (TLV's) (see Chapter 10) is the use of 1/30th of the TLV; the logic is that about 1/3rd (5/14ths) of the air intake is at work and that the public near a plant should not be exposed to more than 1/10th of an employee's exposure since they are not paid to accept risks and they represent a larger fraction of the population than the industrial group. This procedure has been used for ionizing radiation limits around nuclear facilities, 170 mrem versus 5 rem and for sulfur dioxide level in Britain, 0.17 ppm_V versus 5 ppm_V. Taking 1/30th of the TLV's yields such values as 330 $\mu g/m^3$ for nuisance particulates, 0.003 ppm_V for ozone, 33 $\mu g/m^3$ for sulfuric acid, 1.7 ppm_V for carbon monoxide, 0.17 ppm_V for nitrogen dioxide, and 0.33 ppm_V for hydrogen sulfide. It may be noted that some of these values are unrealistic.

Danger points for the common pollutants have been defined by the Environmental Protection Agency (EPA) to guide the States in setting their control strategies for emergency episodes of air pollution. Levels of the pollutants that could cause significant harm to human health are:

i. Sulfur dioxide--2620 $\mu g/m^3$ (1 ppm_V) for 24 hr;

ii. Particulate matter--1000 $\mu g/m^3$ or 8 Cohs* for 24 hr;

iii. Sulfur dioxide and particulate matter--490,000 $(\mu g/m^3)^2$ for the 24-hr product of concentrations or 1.5 ppm_V-Cohs for 24 hr.

iv. Carbon monoxide--57.5 mg/m^3 (50 ppm_V) for 8 hr, 86.3 mg/m^3 (75 ppm_V) for 4 hr, or 144 mg/m^3 (125 ppm_V) for 1 hr.

v. Photochemical oxidants--800 $\mu g/m^3$ (0.4 ppm_V) for 4 hr, 1200 $\mu g/m^3$ (0.6 ppm_V) for 2 hr, or 1400 $\mu g/m^3$ (0.7 ppm_V) for 1 hr.

vi. Nitrogen dioxide--3750 $\mu g/m^3$ (2 ppm_V) for 1 hr or 938 $\mu g/m^3$ (0.5 ppm_V) for 24 hr.

---

*Coefficient of haze (Cohs) = 10^5 X Absorbance/Linear feet sample.

V. EMISSION LIMITS

After the standards for a region are set, emissions limits which will meet these standards must be formulated. If the standards were set with due cognizance of their intended purpose and compatible with feasible cleanup capabilities, the drawing up of emissions limits will not be overly difficult. Emissions limits are instituted such that the ambient quality will be at least as good over the entire region as that specified in the standards. The formulation of emissions limits is accomplished by considering the additive effects for all the sources within the region plus the pollutant coming into the region from adjacent regions. The number, size, and location of the sources and the atmospheric dispersion and natural decay of the pollutant are used to calculate the resultant ambient concentrations probable at various points in the region. Computer mapping techniques permit the determination of the "hot spots," spots with the highest ambient concentrations expected. For example, the area is often divided into a grid system; the concentrations expected for various wind and dispersive conditions are determined for each grid point; and an isodose plot is made to reveal the high concentration areas. The methods of calculation are discussed in Chapters 3 and 14.

The ultimate in setting emissions limits would be to set the limits for each source individually; however, the regulatory process has not reached such a point of sophistication at the present time and about the best that can be done is to set uniform emissions limits at whatever the strictest required limit for the region figures to be. The current procedure makes for easier enforcement of the emissions limits. It is this easier enforcement aspect that has led many areas to abandon land-use concepts in setting their standards.

Federal emissions limits have been set for some of the major stationary air pollution sources (see Table 11-2) and have been planned for many others. Controls on toxic materials emissions will also be set by the Federal authorities. The first ones that have been issued are for beryllium, mercury, and asbestos. The principal features of the auto exhaust regulations that have been federally legislated are found in Table 24-2.

TABLE 11-2

Federal Emission Limits for New Stationary Sources[a]

Source	Pollutant	Limit[b]	
Fossil fuel-fired steam generators	Particulates	0.1	$lb/10^6$ Btu input
	Plume opacity	20%	(40% 2 min any hr)
	Sulfur dioxide		
	Oil	0.8	$lb/10^6$ Btu input
	Coal	1.2	$lb/10^6$ Btu input
	Nitrogen oxides[c]		
	Gas	0.2	$lb/10^6$ Btu input
	Oil	0.3	$lb/10^6$ Btu input
	Coal	0.7	$lb/10^6$ Btu input
Incinerators	Particulates	0.08	gr/scf @ 12%$_v$ carbon dioxide
Portland cement plants	Kiln		
	Particulates	0.3	lb/ton kiln feed
	Plume opacity	10%	
	Clinker cooler		
	Particulates	0.1	lb/ton kiln feed
	Plume opacity	10%	
Nitric acid plants	Nitrogen oxides	3	lb/ton acid
	Plume opacity	10%	
Sulfuric acid plants	Sulfur dioxide	4	lb/ton acid
	Acid mist	0.15	lb/ton acid
	Plume opacity	10%	

[a]"Rules and Regulations," Federal Register, 36:247, 24876, December 23, 1971.

[b]Maximum 2-hr average.

[c]In ppm$_v$ (dry) @ 3% excess oxygen, 175, 230, and 500.

VI. SUMMARY

The chosen course of action for setting emissions limits in this country is the writing of criteria, the determination of goals, the delineation of standards, and the regulation of the emissions. Criteria summarizing the effects of the pollutant when present at various concentrations have been written for the commonly considered pollutants. Goals should be set by an informed citizenry of the region and regulated with consideration of the best available information on costs and benefits. The goals are then used by the regulatory agency for setting standards. Finally, emissions limits are set to achieve the standards over the entire region.

Criteria should summarize the effects--health, plant damage, material, and esthetic--which have been reported from reliable studies. Because of the variability encountered in the various effects levels, statistics plays an important role in any reliable study. The probable health effects should be based on epidemiological and toxicological data. The levels reported by the two techniques will vary because of the inherent differences in size of population, length of exposure, and other factors. For example, a toxicological study shows beneficial effects from a 1-year exposure of guinea pigs to 5 ppm_v of sulfur dioxide, whereas epidemiological studies show increases in human morbidity and mortality from levels of a few hundredths of a ppm_v. Both sets of data may be reliable and must be interpreted with the philosophy of the toxicological-epidemiological dilemma.

The goals should be set by the local people of the region. They should be well informed by technical people of what various levels mean to them in terms of their own values--what the visibility will be for certain suspended particulate concentrations; how often they will have to repaint their houses because of hydrogen sulfide damage; how much they will notice the odors of hydrogen

sulfide and sulfur dioxide; and what other cause-effect relationships they can expect. The people also need to be told what
levels can be attained by what means and at what cost. If they
are properly informed about the situation, they can help set meaningful goals which have some possibility of being met.

The regulatory agency will have technical personnel who
are capable of translating the goals for the region into standards
and emissions limits. The standards must be discussed with the
goal setters to show them when and how the goals are to be met.
The standards must be achievable within a period of not more than
a few years; otherwise, the standards will bog down in a morass
of variances and unenforced emissions limits

The ideal management of air pollution must have the proper
balance among sociological desirability, technological capability,
and economic feasibility.

PROBLEMS

1. What should be found in a good set of criteria for alfalfa
   damage by sulfur dioxide? For gladioli damage by ozone?
   For citrus fruit damage by photochemical oxidant?

2. Find examples in the literature which support the third part
   (small doses stimulate) of the Arndt-Schulz Law.

3. Compare and contrast the Air Quality Criteria for Photochemical
   Oxidants (ozone) and the "Community Air Quality Guide" of
   AIHA for ozone. Why the differences?

4. What ambient air quality goals have been set for your region?

5. Check the compatibility of the 60 μg/m^3 annual geometric
   mean for suspended particulate matter and the 125 μg/m^3
   24-hr, 1 percentile value by the methods of Larsen.

6. What hydrogen sulfide level would be permitted by the 1/30th
   TLV rule of thumb? How does this compare with the Odor
   Threshold for hydrogen sulfide?

BIBLIOGRAPHY

The bibliography of Chapter 10 is pertinent to this chapter and reference is made to it, but the listings will not be repeated.

Air Quality Criteria for Carbon Monoxide, National Air Pollution Control Administration, Publ. No. AP-62, Washington, D. C., March 1970.

Air Quality Criteria for Hydrocarbons, National Air Pollution Control Administration, Publ. No. AP-64, Washington, D. C., March 1970.

Air Quality Criteria for Nitrogen Oxides, Environmental Protection Agency, Publ. No. AP-84, Washington, D. C., January 1971.

Air Quality Criteria for Particulate Matter, National Air Pollution Control Administration, Publ. No. AP-49, Washington, D. C., January 1969.

Air Quality Criteria for Photochemical Oxidants, National Air Pollution Control Administration, Publ. No. AP-63, Washington, D. C., March 1970.

Air Quality Criteria for Sulfur Oxides, National Air Pollution Control Administration, Publ. No. AP-50, Washington, D. C., January 1969.

Cassell, E. J., "Are We Ready for Ambient Air Quality Standards?" J. Air Pollution Control Assoc., 18:12, 799-802, December 1968.

Chapman, R. L., "Present Status of Air Quality Criteria," Beckman Bulletin, May 1968, 5 pp.

"Community Air Quality Guides," Am. Industrial Hyg. Assoc. J.,
"Aldehydes," 29:5, 505-512, Sept-Oct 1968.
"Beryllium," 29:2, 189-192, Mar-Apr 1968.
"Carbon Monoxide," 30:3, 322-325, May-June 1969.
"Ethylene," 29:6, 627-631, Nov-Dec 1968.
"Inorganic Fluorides," 30:1, 98-101, Jan-Feb 1969.
"Iron Oxide," 29:1, 4-6, Jan-Feb 1968.
"Lead," 30:1, 95-97, Jan-Feb 1969.
"Ozone (Photochemical Oxidant), 29:3, 299-303, May-June 1968.
"Phenol and Cresol," 30:4, 425-428, July-Aug 1969.
"Rationale," 29:1, 1-3, Jan-Feb 1968.
"Sulfur Compounds," 31:2, 253-260, Mar-Apr 1970.
"Total Particulate Matter," 30:4, 428-434, July-Aug 1969.

Larsen, R. I., "Relating Air Pollutant Effects to Concentration and Control," J. Air Pollution Control Assoc., 20:4, 214-225, April 1970.

Ledbetter, J. O., "Statement on the Proposed Ambient Air Quality Standards for the Dallas-Fort Worth Air Quality Control Region," Prepared for Texas Chemical Council, Texas Air Control Board Hearing, Dallas, Texas, July 1970.

Patty, F. A., Ed., Industrial Hygiene and Toxicology, Vol. II: Toxicology, 2nd rev. ed., Interscience Publishers, New York, 1962.

"State Officials Speak Out on Air Quality Standards," Environmental Sci. Tech., 4:1, 21-23, January 1970.

Chapter 12

BASES FOR AIR POLLUTION ABATEMENT DECISIONS

I. INTRODUCTION

Engineering decisions for air pollution abatement should be based on the following: what the problem is and its magnitude; how alleviation of the problem can be achieved; what will the control cost; and does company policy dictate other than the least cost control. This chapter delineates the items that need consideration for the usual case decisions under the topics of characteristics of the gas stream, efficiency required, estimating cost, other useful design information, selection of abatement method, and proving abatement method. It is specifically intended as a prod to the thinking and not as a substitute for good engineering judgment, which is the very best basis for engineering decisions. Decision theory and operations research are not covered, but rather the parameters which would need to be input to such calculations are described.

II. CHARACTERISTICS OF THE GAS STREAMS

Any satisfactory control of air pollution requires some knowledge of the characteristics of the polluted gas stream. Some characteristic parameters are always needed (flow rate, temperature, and type of pollutant); some additional parameters are usually needed; and quite complete characterization is sometimes required. Figure 12-1 shows the position of the gas stream characteristics in the overall scheme of air pollution control selection.

In most instances the gas stream can be considered as air for design calculations involving such parameters as density, viscosity, and specific heat, but for the very high moisture con-

Primary basis: emission limits/ambient air standards
Secondary basis: public relations, employee morale, probable future regulations

Penetration allowed/efficiency required

    Gas stream characteristics
      Volumetric flow rate and fluctuations
      Temperature
      Pollutant properties
        Chemical identity
        Concentration/grain loading
        Solubility
        Combustibility
        Hazards--health, fire/explosion
        Plus for particles
          Size
          Density
          Shape
          Abrasiveness
          Electrical properties
            Dielectric constant
            Bulk resistivity
          Handling properties
            Bulking/caking
            Hygroscopicity
      Moisture content
      Dew point
      Density
      Viscosity
      Total chemical composition
      Odors

Abatement method

    Gas cleaning

Gaseous pollutants	Particulate pollutants
Absorption	Filtration
Adsorption	Electrostatic precipitation
Combustion	Scrubbing
Condensation	Inertial collection

             Others/combinations

    Prevent formation/release

    Environmental dispersion for natural decay

Start-up tests/adjustments

Acceptance/provisional acceptance/rejection

Cost factors

First	Continuing
Engineering	Power
Hardware	Materials disposal
Installation	Maintenance
Auxiliaries	Interest

Facilities required
    Space
    Water
    Other utilities
    Heat recovery
    Combustion air

Company policy
    Buying
      Major manufacturer
      Proved equipment
    Worth of money
    Minimum care system

FIG. 12-1. Selection of abatement method.

tents and a few other instances with wide variations, the composition differs markedly from that of air, and cognizance must be
made of this fact. Pollutant concentrations normally range from
a few ppm_v to a few thousand ppm_v, and the stream will not
differ greatly from air in its composition. Combustion will de-
plete the oxygen and add carbon dioxide in percentage amounts
with only small effects on many of the design parameters because
the two gases have similar thermodynamic properties. Combustion
of hydrogen-rich fuels adds significantly to the moisture of the
flue gas.

A. Always Required

Some properties of the polluted gas stream must be known
regardless of the abatement method to be used. The most important single factor is the volumetric flow rate (Q). There are
widespread ambiguities in the listed Q values. Probably the
best method for reporting flow rate is a simple listing of the cfm
with its temperature and water content, at least when water makes
up a significant portion of the total flow. For clarity the actual
cfm is abbreviated acfm. When acfm is corrected for temperature
(to 20°C) and pressure (to 1 atm), it becomes the normal cfm or
Ncfm; but unfortunately, in the United States the Ncfm is usually
called standard cfm or scfm. Sometimes the volume is corrected
for the moisture content; if this is done, the value should be
identified as scfm(dry) or Ncfm(dry). Moisture in the ambient
air will not exceed a few percent even at saturation (e.g., $2.31\%_v$
at 20°C); however, in the hot gases of ducts and stacks the
moisture content may be a major portion of the total flow.

The temperature is a vital measure, even for values reported
as scfm. The designer needs to know the temperature in order to
use construction materials that will withstand the temperature,
and there is a need to know the temperature because the density
and the viscosity of the gas stream depend on the temperature
and they are often required for the engineering design of equipment.

In addition to the volumetric flow and the temperature, only
the pollutant type and concentration are needed in virtually all
instances. For gaseous components, the concentrations are
usually given in parts per million by volume (ppm_v). The reason
for using ppm_v instead of percent is that the concentration is
usually rather dilute and $\%_v$ would be a very small number. For
example, the 330 ppm_v of carbon dioxide in ambient air is only
$0.033\%_v$. Particulate loadings are normally given in grains per
standard cubic foot (gr/scf) or per actual cubic foot (gr/acf). A
grain is 1/7000th of a pound. Grain loadings usually range from
a few hundredths of a gr/acf to several gr/acf. A grain loading
of 10 will usually require more than 99% removal by current regulations; in pneumatic carry of solid materials, grain loadings of
hundreds may be achieved. An effort is being made by some
people in regulatory agencies to list all pollution concentrations,
both gaseous and particulate, in $\mu g/m^3$ for ambient and mg/m^3
for stacks and flues.

The "type of pollutant" as used above normally means the
generic identification for gaseous and sometimes for particulate
pollutants; however, for inert particles and some vapors, control
may hinge not on an exact identification but simply on the properties that affect collection, dispersion, or control.

B. Usually Required

For the usual problem, gas cleaning is the required solution,
and gas cleaning design needs some properties in addition to the
basic ones listed above. Some of these characteristics may not
be needed for more than one control method, but they are required
in order to permit consideration of all possible control methods.
These parameters often include flow rate fluctuations, dew point,
moisture content, density, viscosity, corrosiveness, and hazards
(fire/health) presented by the total gas stream plus certain characteristics of the pollutant which affect treatability. The latter

include the following factors: solubility, combustibility, and
adsorptivity for gaseous pollutants; and size, density, shape and
abrasiveness, solubility, combustibility, electrical properties,
and handling characteristics for particulate pollutants.

Flow fluctuations have a strong influence on the collection
efficiency of many cleaners and a significant effect on most effective stack height calculations. As a result, the size and
frequency of such fluctuations need careful consideration in the
design process. The moisture content affects the volume calculations, density, and other gas stream parameters. The dew
point is required for stack design and most control design and it
is vital for fabric filter installations. Measurement of the dew
point is advisable despite the difficulty involved (see Chapter 9)
because calculation of the dew point from the moisture content
can be grossly in error if sulfurous gases are present.

The corrosiveness of the gas stream and the collected material influence the choice of design materials and temperatures.
Some fairly simple tests on existing installations or a literature
description of such tests or corrosion problems can guide design
decisions.

The hazards to health and of fire/explosion in the gas
stream and the collected materials must be known in order to
avoid problems in these areas. The percent of the lower explosive
limit should be narrowly defined for gas streams with highly flammable constituents. Changes which are likely to occur in concentration and any tendency of the flammable component to condense
out of the gas stream are particularly important. For combustion
calculations, the heating value of the gas stream should be known
so that fuel costs can be calculated.

The treatability and subsequent disposal of the polluted
gas stream by some methods is dependent on the pollutant solubility, combustibility, and adsorptivity. The first two factors may

be needed for either gaseous or particulate pollutants and the
last for gaseous pollutants.

For particulate pollutants a rather good size distribution is
usually needed. The importance of the particle size to the calculations of collection efficiencies and plume opacities and to
estimates of atmospheric residence times needs strong emphasis.
The shape and abrasiveness of particles will determine whether
a cyclone or other high velocity equipment would need a rubber
or plastic liner and how often filter bags would likely need replacing.

The electrical properties of particles are needed for electrostatic precipitator design and may play some role in judgments
on other collection devices such as filters and inertial cleaners
and on particle handling, but these latter relations have not been
quantified.

The handling properties and nature of collected particles
are quite important in control equipment selection. Bulking or
caking by collected particles can cause havoc on fabric filters
or in hoppers of collecting equipment. The fact that collected
materials must be disposed of in some manner or returned to the
process plays an important role in the selection of control equipment, especially wet collection device feasibility.

The chemical composition of particles is often quite helpful.
The composition is not always obvious; e.g., the dust from a
cement kiln is not necessarily cement. A large fraction of such
dust has not reached the temperature necessary for calcination.

C. Sometimes Required

Rarely will the total composition of the gas stream be needed
for abatement, but such information, when it is available, should
be furnished to the design engineer. Also, certain installations
will require special data that are peculiar to that type of installation, especially radioactive wastes.

Odors need 'to be identified as nearly as possible and their concentrations defined. Some odors will be characteristic of the emitting process, e.g., sulfide odors from kraft paper and acrolein from rendering.

Molecular diffusivity might be needed for the design of absorption or adsorption removal of a gaseous pollutant or for design calculations relating to the injection of a polluted gas stream into an abandoned mine or other application that removes the particles by molecular impaction.

III. ESTIMATING COSTS

Effective planning requires that estimates of cost be made for air pollution abatement--prevention or control. If two or more types of abatement methods will do the job, selection of the process may well be made on the basis of the cost per year for control by each of the methods. In nearly all cases, including those where the method of abatement is dictated by the type of pollutant and the degree of control required, cost estimates must be determined so that financing can be arranged, product prices can be adjusted to absorb the cost, and decisions on contract letting can be wisely made.

A. Comparison Estimates

Most estimates of cost are made by comparisons. The construction and contracting magazines run cost breakdowns on items in recent contracts. Estimates may be made from use of the company's own known costs for items or from the data of other firms.

1. Using Known Company Costs

An existing plant knows its costs for items such as heat, refrigeration, steam, water, electricity, and other usual operating expenses. Some representative operating costs for such items are shown in Table 12-1. A new plant can often estimate these

AIR POLLUTION

TABLE 12-1

Some Representative Operating Costs

Item	Approximate cost	Remarks
Electricity	$0.01-0.02/kWh	Low range of costs usually apply for large plant in Gulf Coast area
Horsepower	$100/BHP-yr	
Water pumpage	$4/yr-gpm	High range of costs usually apply for small plant in Northeast, Great Basin, and interior Southwest areas
Water		
Process	$0.05-0.30/1000 gal	
Cooling	$0.01-0.03/1000 gal	
Sewer	$0.10/1000 gal	
Fuel	$0.15-1.25/10^6 Btu	Low for coal or lignite at plant. High for gas with long transport
Steam	$0.30-2.00/1000 lb	
Refrigeration	$0.40-1/std ton-day	

costs fairly well by knowing the contract prices of the available
utilities.

2. Using Cost Data of Others

It is a good and often followed engineering practice to
estimate costs for an air pollution control installation on the
basis of what a similar recent installation cost. The availability
of such data should not be taken as the easy solution to making
a cost estimate. Several uniqueness factors may change the
costs from one plant to another, including the cost variance of
such items as site clearance, foundation, freight, utilities,
disposal of collected material, and local labor. Cognizance
should be taken of all possible factors in adjusting the estimate
for differences in location, operations, and inflationary trends.
The use of cost extensions on individual items provides the flexibility needed for varying design parameters and quantities.

B. Rules of Thumb

Some useful rules of thumb can give rough estimates of
cost with a minimum of effort. One such quite common rule is

that equipment will cost $1 per pound of weight. This number
will usually give an estimate which is within a factor of 2 in the
majority of cases and is much closer in some instances. Economy
of scale is often estimated by the six-tenths rule (see Fig. 12-2);
the calculations are made using the formula

$$C/C_0 = (S/S_0)^{0.6},$$ (12-1)

where C = cost of equipment with size S and C_0 = cost of same
equipment with size S_0. Sometimes a power of 0.7 is used with
Eq. 12-1, especially for total plant costs. Economy of scale
obviously does not hold in all cases; e.g., some items have a
fixed price regardless of the number bought and the cost per foot
for tall stacks goes up, not down, with height.

For many air-cleaning processes, including electrostatic

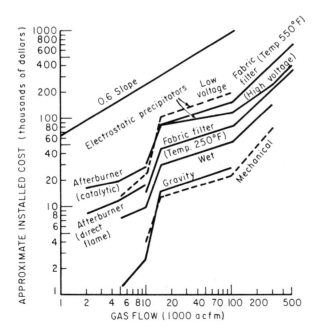

FIG. 12-2. Cost versus size for particulate collectors.
From "Particulate Controls: A Must to Meet Air Quality Standards,"
Env. Sci. & Tech., 3:11, 1149, November 1969.

precipitators, absorbers, spray towers, and other time-rate
processes, the efficiency of removal is related to the retention
time of the gas stream in the cleaner; therefore, the efficiency
is related to the size of the cleaner. An estimate of the cost for
lowering the penetration (raising the efficiency) can be made by
assuming

$$q = A/S, \tag{12-2}$$

where q = fractional penetration of pollutant through the cleaner
= one minus fractional efficiency and A = a constant for a given
cleaner. This is equivalent to assuming that the negative exponential curve can be represented satisfactorily by a hyperbolic
function. Then,

$$q_0/q = S/S_0 = (C/C_0)^{3/2} \text{ or } C/C_0 = (q_0/q)^{2/3}. \tag{12-3}$$

There are, of course, exceptions to this generality, the most
notable being bag filters, combustion units, and adsorbers.
These latter types of cleaners have efficiencies that are only
slightly related to the size of the cleaner as long as the size
of the cleaner is adequate. Similar logic might be applied to
other cleaners; for example, the efficiencies of some inertial
cleaners are related to the power used in the cleaner rather than
to the size of the cleaner.

The installed cost of air pollution control equipment will
usually run 2 1/2 to 3 1/2 times the base price of the equipment;
therefore, the installed cost may be taken as 300% of the purchase
price for an estimate which is at least good for preliminary purposes. The breakdown is approximately as follows: Equipment
Cost 100% + Installation 100% + Auxiliary Equipment 100%.
"Loading" items on installation total about 60% of the purchase
price; these items include sales tax, engineering, and contractor's
profit and overhead.

The use of expensive materials in control equipment is reflected in the cost of the equipment. Calvert (see Stern reference)

cites the following relative costs for absorption towers: carbon
steel 1.0 (basis), copper 1.4, aluminum 1.5, lead 1.6, #304
stainless steel 2.3, #316 stainless steel 2.7, Monel or nickel
3.0, and Hastelloy 3.5. The prices of these metals fluctuate
and such fluctuations will change the ratios presented.

C. Calculating Annual Costs

The best basis for comparing costs of air pollution control
is the total annual cost. This cost includes the capital, operating,
and maintenance costs. Annual capital costs are calculated by

$$C_a = \left(\frac{P}{N}\right) \frac{(1 + i)^n}{(1 + i)^n - 1} , \qquad (12\text{-}4)$$

where C_a = annual capital costs, P = initial installed cost of
equipment, N = lifetime or amortization base for equipment (yr),
i = compound interest cost for money, and n = number of interest
periods.

EXAMPLE 12-1: What is the annual capital cost for air pollution
   control equipment with an installed cost of $300,000 that
   is amortized over 10 yr at 10%?

$$C_a = \left(\frac{\$300,000}{10}\right) \frac{(1 + 0.10)^{10}}{(1 + 0.10)^{10} - 1} = \$48,823.62$$

Taxes may be included as part of the capital costs, may be a
separate item, or may be included in operating costs according
to the policy of the particular company on its bookkeeping procedures and the tax laws.

1. Operating Costs

The operating costs include power, labor, materials,
utilities, disposal of collected materials, and absorbent, adsorbent, and catalyst renewals. Operating costs should include all
direct costs incurred after installation except maintenance.

Power costs are the fan costs to move the air against the head loss in the equipment and ducts and the power to run the equipment and its auxiliary equipment. The power requirements for fans may be calculated using the equation

$$HP = Qh/(6356\ E),\qquad\qquad (12\text{-}5)$$

where HP = brake horsepower, Q = air flow rate (cfm), h = head loss (in. w.g.), and E = mechanical efficiency of fan system.

EXAMPLE 12-2: What is the annual cost of moving 1000 cfm of air against a head of 1 in. w.g. if electricity costs are $0.015/kWh and the efficiency is 60%?

HP = 1000 X 1/(6356 X 0.60) = 0.262

0.262 HP (8760 hr/yr) (0.746 kWh/HP-hr) ($0.015/kWh)
    = $25.70

It should be kept in mind that large air flows and large head losses result in quite large power costs; e.g., a venturi scrubber may have a capacity in hundreds of thousands of cubic feet per minute and a head loss of more than 60 in. w.g.

Disposal of collected material will usually cost 10-25% as much as the collection of the material but may be even higher. The collected material must be rather valuable for the air cleaning to pay its own way.

2. Maintenance Costs

Virtually all equipment requires some maintenance during its lifetime. The divide between operating and maintenance costs is not clearly drawn in all cases, and it does not need to be for engineering purposes as long as the expenditures are included in one place or the other. In addition to the routine maintenance of replacing worn out belts, lubricating, painting, and such, there is normally some unforeseen expense for breakdowns, sampling of effluents, etc.

D. Inflation/Deflation Adjustments

Changes in the economy are reflected in air pollution control costs. Perhaps the best estimate of changing prices can be
made by plotting one or more of the cost indexes for a considerable
period of immediate history, then extrapolating the future trend of
the index. Indexes of this type are published by the Department
of Commerce, Engineering News Record (ENR), Chemical Engineering (Marshall and Stevens Index), and others. Each index has
some past time as the selected base; e.g., the Marshall and
Stevens Index (MSI) for chemical equipment uses 1926 costs for
a base of 100. The MSI has run 238, 239, 242, 245, 253, 263,
273, 285, 303, 321, and 332 for the years 1962-1972, respectively. The construction cost index of ENR is based on 1913 and
showed averages near 1600 for September 1971 versus about 1100
for September 1967. Chemical Engineering also publishes a Plant
Cost Index with a 1957-1959 base of 100; it was 137 for 1972.
The extrapolation of a plot for future estimates should be guided
by added constraints when they are known.

E. Contingencies

Every estimate of costs should contain a line item for contingencies. This item is often set at 10% for rather detailed
estimates, but may range up to 100% or more for rough estimates
and/or long-range planning. An estimate without such an item
should be regarded as suspect--either the uncertainty has been
distributed over the other items or it has been omitted.

IV. SELECTION OF ABATEMENT METHOD

The method chosen to abate an air pollution problem depends
on many factors--some factors are common to nearly all problems
while others are unique to the particular plant. The three distinct
categorical types of control for air pollution problems are the
following: (1) the prevention of formation and/or release of the

pollutant in the process; (2) the dispersion of the pollutant to
prevent problem concentrations of the pollutant before natural
decay; and (3) the removal of the pollutant from the process gas
stream. All three types of abatement should be considered, at
least enough to be assured that one or more types of abatement
are not applicable to the particular situation.

A. Practicality of Source Elimination

One selection question to be answered is whether it is
practical to eliminate the pollutant at its source--the prevention
of its formation and/or release (see Chapter 14). This type of
control might be achieved by process change, fuel substitution,
or plant shutdown. The process emitting the pollutant may be
modified to eliminate the formation of the pollutant. In many
instances there are alternative methods for the production process,
and although the process engineer will likely be loath to change
the process, such a change may be the simplest solution.

Many plants, particularly the old inefficient ones, will
find it more economical to cease operation than to institute air
pollution controls. The cost of air pollution control would generally be expected to have a greater effect on the world market
position of an industry than on the domestic market position because the domestic competitors would presumably have to meet
the same regulations.

B. Feasibility of Environmental Dispersion

Although almost everyone involved in air pollution control
will concede that the objective of control is to provide good
quality air for people, animals, and materials and that good
quality is evidenced by low concentrations of pollutants, many
will not admit that environmental dispersion or dilution is good
practice. Regulations based on property line concentrations have
generally given way to regulations on plume opacity and/or
emission rates. Dispersion from stacks is still widely used by

the major sulfur dioxide sources such as smelters and power plants. There are many instances other than high stacks where dispersion is the principal control; this is especially true for the "unusual" pollutants such as noise, ionizing radiations, and water fog. The inverse square law for the energy flux lends enhancement to control by distance, and the relocation of a cooling tower can provide dilution of cold weather "steam fog" before it reaches a highway and causes visibility problems.

EXAMPLE 12-3: What will be the sulfur dioxide concentration (ppm$_v$) in the flue gas from a power plant which burns coal with 2%$_w$ sulfur and adds 140 scf of air per pound of coal? Assume 10%$_w$ of moisture and 4%$_w$ of hydrogen in the coal.

The volume of sulfur dioxide and carbon dioxide produced equals the volume of oxygen used.

Volume of Water as Additional Flue Gas:

$[0.10 \text{ lb} + 1/2 (0.04) (18/2) \text{ lb}] (385 \text{ ft}^3/18 \text{ lb}) = 6.0 \text{ ft}^3$

Concentration of Sulfur Dioxide: $PP = \text{ft}^3/\text{ft}^3$ atm

$$\frac{0.02 \text{ lb S/lb coal} (2 \text{ lb SO}_2/\text{lb S}) (385 \text{ ft}^3/64 \text{ lb}) (10^6 \text{ppm}_v)}{140 + 6.0 \text{ ft}^3/\text{lb coal}}$$

$= 1650 \text{ ppm}_v$

EXAMPLE 12-4: What is the estimated Ringelmann number for a fossil fuel power plant plume that is 18 ft in diameter and has 0.10 gr/acf of fly ash? Assume that the particle characteristics give a β (extinction coefficient) of 0.18 ft^3/gr-ft.

Ringelmann No. = % obscuration/20

% obscuration = 100 - transmittance = $100 - I/I_0$

$I/I_0 = \exp(-\beta cL)$ Beer-Lambert Law (see Chap. 7)

$I/I_0 = \exp[-0.18 (0.10)(18)] = \exp(-0.324) = 0.723$ or 72%

Ringelmann No. = $(100 - 72)/20 = 1.4$

Lifetimes of pollutants in the air are not well known, even for the common pollutants. In order to make the best engineering design for environmental dispersion, the best estimates of decay rates must be used; however, adequate problem prevention on a local basis can usually be designed by knowing the dilution factor required; i.e., the stack concentration divided by the maximum

allowable ground concentration. The additivity of sources must
be taken into account (see Chapter 15). The decay rates for the
common pollutants should be adequate to prevent large area problems if local problems are prevented. For example, the average
lifetime of sulfur dioxide at usual relative humidities of ambient
air appears to be about 4 hr.

C. Requirement of Gas Cleaning

Most air pollution problems require gas cleaning; the polluted gas stream is passed through a collector to remove the of-
fending pollutant(s) or through a process that converts (usually
by oxidation/reduction) the pollutants to innocuous products.
The degree of cleaning required is usually determined by emissions
limits, and the cleaning process is selected on an economic basis.
Sometimes, there are other factors which cause treatment design
efficiency to be higher than would be required by the emissions
limits; these are the probability that the emissions limits will be
lowered during the lifetime of the plant or esthetic concerns for
better public relations or employee morale. Often the lower of
two regulations must be met, for example, grain loading and plume
opacity. The method selected may be other than the one which
shows the lowest estimated annual cost because of considerations
such as proved application, company uniqueness, and operating
difficulty attached to the economical process.

When gas cleaning is required, the recovered product will
seldom pay for the cleaning. Air pollution abatement measures
are sometimes given short shrift because they are "non-pay" items.
Such an attitude on the part of plant management can ultimately
result in excessive costs for the abatement procedures. One of
the principal drawbacks of such an attitude is that the procrasti-
nation will lead to insufficient time to make the proper design and
installation in an unhurried manner.

1. Penetration versus Efficiency

In most air pollution abatement work, the use of penetration
(q), the fraction of the inlet concentration that remains after the

cleaning process, rather than the collection efficiency (η), the fraction of the inlet concentration that is removed by cleaning, simplifies matters. Logic indicates that what is taken out is of lesser importance that what is left. Admittedly, there is a relationship between the two; namely,

$$\eta + q = 1.00 = 100\%. \tag{12-6}$$

If cleaning were done on the basis of collection efficiency, a power plant which burned higher ash coal could be favored over one with lower ash coal, and the efficiency might be met by adding large, easily collected material to the stack gas.

2. Efficiency Required

The overriding factor in gas cleaning equipment selection is the efficiency required, or the penetration allowed. Simply setting the design efficiency normally narrows the selection of cleaning equipment to no more than a few types. The q plays a twofold role in abatement selection; i.e., whether the equipment will meet the q and what it costs to get the required q with the particular equipment.

a. Gaseous Pollutants. The primary methods available for the cleaning of gaseous air pollutants are absorption, adsorption, and combustion (oxidation/reduction). The methods of lesser importance include condensation, reaction, odor masking or counteraction, and storage for chemical or radioactive decay. The control of highly hazardous gaseous materials is likely to be combustion when oxidation renders the substances harmless because of the reliability and simplicity of incineration. Very malodorous materials are usually burned with flame or catalytic combustion, or they are adsorbed on activated carbon. Absorption is normally used where the gaseous component is rather soluble and some penetration of the pollutant will not be ruinous to the control effort.

b. Particulate Pollutants. High efficiency can be obtained for very small particle (diameters less than 1 μm) removal with

only a few gas cleaning processes; these are filtration, electrostatic precipitation, and high energy scrubbing.

The particulate cleaners with midrange efficiencies are the
induced-spray, bubble plate, and spray tower scrubbers, cyclonic
scrubbers, and the multiple cyclone or other high centrifugal force
device. This class of cleaner is often sufficient for particles
produced by crushing processes and mists from mechanical actions
on liquids or for applications in collecting catalysts or products
with particle sizes that are controlled to be above the minimum
size (a few micrometers) effectively collectable by these processes.

Historically, the low efficiency collectors (capable only of
large particle removal) have been used in air pollution control,
but it is hardly likely that such cleaners will suffice now in any
air pollution application other than as precleaners for more efficient devices. This category of cleaners includes settling chambers, common cyclones, and baffle cleaners.

Calculating the design efficiency of a particulate cleaner
can get quite involved for those cases where the efficiency of
collection is a function of the particle size. The theoretical
penetration (q_T) is usually determined by a rather simple calculation for a given particle size. For discrete groupings of particle
sizes,

$$q_T = \Sigma_i w_i q_i, \qquad i = 1 \text{ to } N \qquad (12\text{-}7)$$

where q_T = fractional theoretical penetration for entire set of
particles, w_i = fractional weight of total particles in ith size
group, and q_i = fractional theoretical penetration for particles
of the ith size. The q_i is obtained by either of two methods;
namely, the top and bottom sizes of the group are averaged and
q_i calculated on the basis of this average size or the q's for the
top and bottom sizes of the group are averaged to get an average
q_i. The methods will give different q_i values if the q_i varies with
other than the first power of the diameter. Furthermore, the weight
distribution within the size group is not likely to be linear. There-

fore, enough size groupings should be used to define q_i as narrowly as desired, without resorting to exotic averaging techniques for the size groups. The overall efficiency may be determined graphically by a plot such as Fig. 12-3. The average ordinate for the collection or the penetration is obviously that which balances the uncollected area beneath the ordinate with some of the collected area above the ordinate when the ordinate and abscissa are drawn to the same scale. Although Fig. 12-3 is a plot of the

EXAMPLE 12-5: Calculate the overall efficiency of collection by Eq. 12-7 and 12-6 for the following conditions (same data as Fig. 12-3):

Size: $43\%_w > 5\,\mu m$; $32\%_w$, 4-$5\,\mu m$; $15\%_w$, 3-$4\,\mu m$;
$6.3\%_w$, 2-$3\,\mu m$; $3.7\%_w < 2\,um$

q_i Values: 0%, $5\,um$; 1.8%, $4\,um$; 13.6%, $3\,um$; 37%, $2\,um$;
100%, $0\,um$

$q_T = 0.43\,(0) + 0.32\,(0 + 1.8)/2 + 0.15\,(1.8 + 13.6)/2$
$+ 0.063\,(13.6 + 37)/2 + 0.037\,(37 + 100)/2 = 5.56\%_w$

$\eta = 100\% - 5.56\% = 94.44\%$

same data used in Example 12-5, the graphical method indicates a higher collection efficiency than does the algebraic method; connecting the points on the graph with straight lines helps in visualizing why this is so.

For the continuous size distribution,

$$q_T = \int \Phi(x)\,q(x)\,dx, \qquad (12\text{-}8)$$

where $\Phi(x)$ = weight distribution of particles with diameter x and $q(x)$ = penetration of particles with diameter x. For the log normal distribution of x, t in the normal distribution function

$$\Phi(t) = 1/(\sqrt{2\pi}\,\sigma)\,\exp(-t^2/2)\,dt, \qquad (12\text{-}9)$$

is represented by $t = (\ln x - \ln \overline{x_g})/\ln \sigma_g$ and the σ by σ_g, where t = normally distributed variate, x = diameter of particle, $\overline{x_g}$ = geometric mean diameter of x (mass median diameter), and σ_g = geometric standard deviation of x. Therefore,

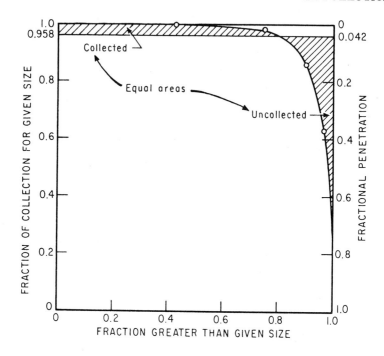

FIG. 12-3. Overall collection efficiency.

$$x = \overline{x_g} \exp(t \ln \sigma_g).$$ (12-10)

The penetration of particles collected by migration is usually calculated on the basis of laminar air flow, a condition that does not obtain in air-cleaning equipment. If turbulent flow is assumed, the probability of collection is random and the penetration may be formulated as

$$q_T = n/n_0 = \exp(-\omega t/s),$$ (12-11)

where n = number of particles remaining airborne after time t, n_0 = number of particles airborne when $t = 0$, ω = migration velocity for particles, and s = distance of migration for collection. For laminar flow, $t/s = 1$ would give $n = 0$, but for turbulent flow n is 37% of n_0. The exponential relationships in efficiencies of collection have spawned the concept of number of transfer units (N_t), where N_t is defined as

$$q = \exp(-N_t). \tag{12-12}$$

For a given design of collector where the migration velocity depends on the diameter, the penetration varies with the diameter according to

$$q = \exp(-kx^\alpha), \tag{12-13}$$

where $\alpha = 1$ for electrostatic precipitators and high efficiency cyclones $= 2$ for settling chambers and low efficiency centrifugal collectors.

For an electrostatic precipitator that is removing log normally distributed particles, the penetration according to Eq. 12-8 becomes

$$q_T = \frac{1}{\sqrt{2\pi}\, \ell n\, \sigma_g} \int_{-\infty}^{\infty} \exp(-t^2/2)\, \exp[-k\overline{x_g}\, \exp(t\, \ell n\, \sigma_g)]\, dt, \tag{12-14}$$

where all terms are as previously defined, except see Chapter 22 for k. Eq. 12-14 is solved by numerical or graphical integration. Common statistical tables give values for $1/\sqrt{2\pi} \int_0^t \exp(-t^2/2)$. For tables which give limits $-\infty$ to t, subtract values from 0.5000 when t is negative or add tabulated values to 0.5000 when t is positive.

3. Specifications and Guarantees

A competent engineering firm not committed to a certain equipment type or manufacturer should be retained for the implementation of air pollution control. The steps to be taken are as follows: (1) selection of the type of control, (2) writing the specifications, (3) estimating the cost, (4) overseeing the installation, and (5) testing the control effectiveness in meeting the specifications. Many companies have hired equipment manufacturers to solve their problems, apparently in the belief that the engineering of the installation would be free. These companies run a high risk that the equipment will not be the best suited for the particular job; the buyer has often received installations that performed very poorly, sometimes even untried experimental designs.

Performance guarantees are not really guarantees in the sense that one may expect the guarantee to be met or no payment is required. If the equipment comes anywhere close to the specified collection efficiency and was installed with "good" work practice and materials, the purchaser must pay for the installation. Because of payment practices, the contractor has usually been paid more than 90% of the contractual amount by the time that he completes erection of the equipment and long before the equipment performance is tested. The price may be a negotiated settlement, but it is likely to be close to the contract price. This guarantee practice presents a strong reason for using penetration rather than collection efficiency. For example, adding $3\%_w$ to the efficiency of $94\%_w$ to obtain $97\%_w$ in many cases does not reflect the difference in size and cost of the equipment for collection nearly as well as does the halving of the penetration from $6\%_w$ to $3\%_w$ (see Section III,B). In order to help in preventing any unscrupulous corner-cutting bid practices, companies that must negotiate a price for equipment that delivers subpar performance should prepare their arguments around penetration values rather than collection efficiencies. The situation on guarantees is improving in that guarantees of meeting the regulations are being made and the regulations are based on emissions.

Perhaps the best guarantee on air-cleaning installations can be had by including air cleaning in the overall plant contract, i.e., letting a turnkey contract for the plant. The contractor will have much more at stake under this arrangement than he will if he is supplying just the air pollution control equipment.

Performance of air-cleaning equipment should be monitored for all probable operating conditions, especially variations in flow rates and pollutant loadings. Emissions from the cleaner should be checked over a significant time in order to spot efficiency changes with age, a procedure that is being written into many acceptance clauses for equipment; this procedure is quite important for catalytic processes.

The test procedures should indicate whether to accept the air cleaner as is, to accept with provisions or reservations, or to reject the cleaner as not meeting the specifications. The equipment manufacturer often puts an engineer on the job to make adjustments in the cleaning process for obtaining the maximum efficiency from the unit. It is not unheard of for this man to stay with a cleaner for more than a year.

V. SUMMARY

The characteristics of the gas stream must be known before an intelligent selection of an abatement method can be made. The volumetric flow rate and temperature of the gas stream are always needed, as is the type of pollutant. The flow rate should be clearly stated as actual cubic feet per minute at the given temperature. Some parameters usually needed for control selection are flow rate fluctuations, dew point, density, viscosity, corrosiveness, and hazards of the gas stream. The treatability of the pollutant should be known, i.e., solubility, combustibility, and adsorptivity for gaseous pollutants and size, density, shape and abrasiveness, solubility, combustibility, electrical properties, and handling characteristics for particulate pollutants. The gas stream characteristics must be carefully determined, although it is frequently quite expensive to do so. Sometimes the actual chemical composition of the polluted stream should be determined.

Cost estimates for air pollution control may be made by comparisons or by rules of thumb. The costs should be reduced to total annual cost by including capital, operating, and maintenance costs. The cost of disposal of the collected material should not be overlooked in making an estimate; it will be rather high unless the collected material is valuable enough to offset this cost. Any estimated cost should include an item for contingencies, then be updated for inflation or deflation.

The selection of the control method should be the lowest total annual cost method, unless company policy dictates some other choice that will also do the job. The alternatives to gas cleaning should be considered; these alternatives are stopping the pollutant emission at its source or dispersing the pollutant into the atmosphere to prevent problem concentrations.

The penetration (amount of material escaping) permitted will often dictate the choice of cleanup methods. The use of the penetration concept instead of collection efficiency in plans and specifications can prevent much trouble in the acceptance of cleaning equipment as being adequate. The collection efficiency attainable by a control process is a function of the pollutant concentration and, as a rule, pollutant characteristics. When the efficiency, or penetration, is a function of particle size, the integrated efficiency on the total particle distribution must be assayed. This overall efficiency can be determined by approximate summation, graphical plot, or numerical integration of the continuous function.

PROBLEMS

1. Convert a stack concentration of 0.012 gr/acf at $950^{\circ}F$ to gr/Ncf. To mg/Nm^3.

2. Calculate the items in Problem 1 for Ncf(dry) and mg/m^3 (dry) if the gas stream has $17\%_v$ of water vapor.

3. If a 100,000-acfm bag house costs \$150,000, what should a 70,000-acfm bag house for the same application cost?

4. Determine algebraically by Eq. 12-7 the probable overall efficiency (η_T) of a cyclone on the dust in Ex. 12-5 if the cyclone can collect $80\%_w$ of the 5-μm particles.

5. Solve Problem 4 graphically.

6. Solve Problem 4 by numerical integration of Eq. 12-13.

BIBLIOGRAPHY

Lund, H. F., Ed., Industrial Pollution Control Handbook, McGraw-Hill, New York, 1971.

Nonhebel, G., Ed., Processes for Air Pollution Control, 2nd ed., CRC Press, Cleveland, Ohio, 1972.

Ross, R. D., Ed., Air Pollution and Industry, Van Nostrand Reinhold, New York, 1972.

Stern, A. C., Ed., Air Pollution--Vol. III: Sources of Air Pollution and Their Control, 2nd ed., Academic Press, New York, 1968.

Chapter 13

AUXILIARY OPERATIONS

I. INTRODUCTION

The total air pollution control system includes capturing, transporting, and cleaning or disposing of the polluted air. Air cleaning is often thought of as merely the collection phase in which the pollutant is removed from the airstream, but auxiliary operations must not be neglected--they are a vital part of air pollution control. The topics that are given brief coverage in this chapter include local exhaust systems, conditioning of airstreams, handling and disposal of collected material, and controls and instrumentation for the air-cleaning process.

II. LOCAL EXHAUST SYSTEMS

The local exhaust system is the air sewer which collects and transports the polluted gas stream to the collector and the cleaned air to the fan and then to the atmosphere. The system plays a major role in the success or failure of air pollution control efforts and therefore merits careful design, construction, and operation. The overriding philosophy of design should be to gather all the pollutant, and the secondary concern should be to keep the volume of polluted air to a minimum because the cost of control depends on the volume to be cleaned.

A. Exhaust Hoods

In keeping with the design philosophy, hoods should be so placed and shaped as to collect all the pollutant; they should have the minimum opening and face velocity to assure efficient process operations and low volumes of polluted airstreams; and,

49

in addition, they should not expose the process operators to the
emissions before the emissions are captured, a factor that is
often overlooked in setting canopy hoods over hot processes.
Not only should the hood come as close as possible to completely
enclosing the process within the guidelines of required access to
the hooded operation but also, insofar as possible, the hood opening should be located to take advantage of the velocity component
that the emissions may have. The velocity pattern at the hood
must be such that all the velocity vectors of pollutant components
will be overcome by the intake air velocities. The good design
considers the magnitude and direction of air currents resulting
from equipment function and from operational activities as well
as the velocities of the emitted gaseous and particulate materials.

A rough estimate of the air being moved with an enclosed
conveyor system may be made by estimating the area of air movement and the velocity of the movement as related to the conveyor
speed. Drinker and Hatch formulated the amount of air moved by
crushed rock falling in a chute as

$$Q = fW^{1/3}h^{2/3} , \tag{13-1}$$

where Q = air flow (cfm), W = weight of rock moved (lb/min),
h = height of drop (ft), and f = ratio parameter with range of 4.7
to 7.8. The equation is based on the data by Chirico and the
range of the variates was as follows: h, 10-30; w, 1220-6700;
and Q, 300-2570.

The velocity of a blowing air jet is reduced to about 10%
of the exit velocity in a distance of 30 diameters. There is an
extensive coverage of jets in the literature. Jets are not satisfactory for trying to dilute pollutants, but they may be used in
the pickup of pollutants from a process, especially solid materials.

For hot processes, Hemeon reports the air volume to be

$$Q = 29 H^{1/3}A^{2/3}z^{1/3} , \tag{13-2}$$

where Q = air flow (cfm), H = heat flow from hot surface (Btu/min), A = area of hot surface (ft^2), and z = distance from hot surface to hood (ft). The expansion of the rising hot gases is allowed for in a canopy hood design by providing an overhang of 0.4z at each end of the hood opening (see Fig. 13-1).

The deceleration of dynamically projected particles was briefly described in Chapter 5. It does merit mention here that

HOOD TYPE	DESCRIPTION	ASPECT RATIO, W/L	AIR VOLUME
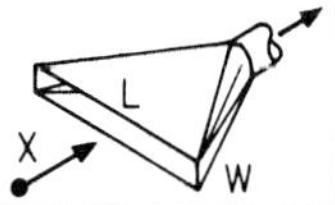	SLOT	0.2 or less	Q = 3.7 LVX
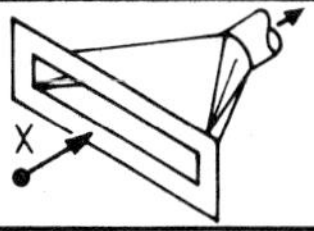	FLANGED SLOT	0.2 or less	Q = 2.8 LVX
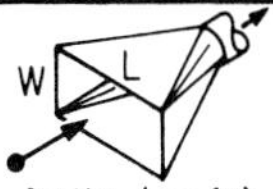	PLAIN OPENING	0.2 or greater to suit work	$Q = V(10X^2 + A)$ $A = WL$
	FLANGED OPENING	0.2 or greater to suit work	$Q = 0.75V(10X^2 + A)$
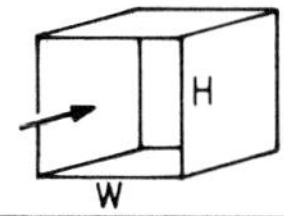	BOOTH	To suit work	Q = VA = VWH
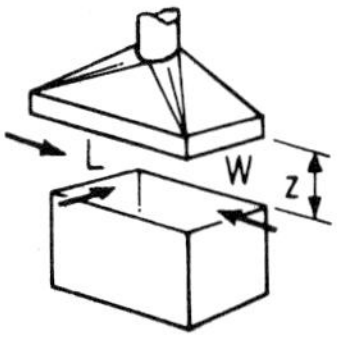	CANOPY	To suit work Normal Overhang = 0.4z	Q = 1.4 PzV P = perimeter of work z = height above work Q = (W+L)zV two sides end losed Q = WzV booth

FIG. 13-1. Hood velocities and volumes. From _Industrial Ventilation_, 10th ed., American Conference of Governmental Industrial Hygienists, Lansing, Michigan, 1968.

the particles of most concern, the respirable particles, are not
likely to travel as much as an inch in air before reaching their
terminal settling velocities.

The velocity pattern around a point sink would have spherical
iso-velocity contours and would be calculated as

$$V = Q/A = Q/(4\pi s^2),\qquad\qquad(13\text{-}3)$$

where V = velocity (fpm), Q = air flow (cfm), and A = cross-
sectional area of sphere around sink at distance s ft (ft^2). This
pattern is altered by the presence of the hood or duct, and the
velocities for application are calculable as shown in Fig. 13-1.
The similarity between the sink formula above and that for round
and rectangular hoods is evident. Note that the benefit of
flanging is reflected in a 25% reduction of the required air volume
to produce a given capture velocity.

Capture velocities often range from 50 to 100 fpm for
evaporative emissions into still air, 100 to 200 fpm for low
inertia emissions such as welding fumes, 200 to 500 fpm for
emissions with moderate inertia into moving air, and 500 to 2000
fpm for high inertia emissions such as particles from grinding or
sandblasting into rapidly moving air. As can be seen by the
factor of 40 from the lowest to highest capture velocities, considerable engineering judgment is involved in selecting a capture
velocity for a specific application. Reference is made to the
bibliographic entries by Brandt and Hemeon and to Industrial
Ventilation for more information about capture velocities.

The head loss of hood entry is important in exhaust system
design. The coefficient of entry for the hood is defined by

$$C_e = Q_a/Q_t = \sqrt{VP/SP},\qquad\qquad(13\text{-}4)$$

where C_e = coefficient of entry, Q_a = actual air flow into hood,
Q_t = theoretical air flow into hood for that static pressure, VP
= actual velocity pressure, and SP = static pressure that produces

VP. The coefficient of entry is relatively unimportant for most
design purposes; what is actually needed is the head loss. The
head loss is related to C_e by

$$h_e = [(1 - C_e^2)/C_e^2] \, VP, \qquad (13-5)$$

where h_e = entrance loss, C_e = coefficient of entry, and VP
= velocity pressure in duct. Total head calculations for the
entire system are frequently carried out in terms of the number
of velocity pressures for the various branches, then converted
to inches of water gage (in. w.g.). Figure 13-2 shows some
common hood shapes and head losses.

B. Ducts

Ducts are conduits to transport the material gathered from
the process for cleaning and exhausting. The velocity in the
ducts must be relatively high for economical purposes, the trade-
off being between the cost of additional duct diameter and pumping
costs of higher head loss. Transport velocities for particles
should be well above reentrainment velocities, maybe twice the
reentrainment velocity. Minimum transport velocities recommended
by Brandt as guides are shown in Table 13-1, or values may be
calculated from the formula

$$V = 6000 \, [S/(S + 1)] \, d_p^{0.4}, \qquad (13-6)$$

where V = transport velocity (fpm), S = specific gravity of particles, and d_p = diameter of particles (in.). A further consideration with combustible mixtures is that a velocity greater than the
flame propagation velocity can add a factor of safety to that provided by flame arresters and dilution to 1/4th of the lower flammable limit.

The head loss in a straight run of duct can be calculated
according to Brandt by

$$L = 0.0365 \, CDQ^{0.09}, \qquad (13-7)$$

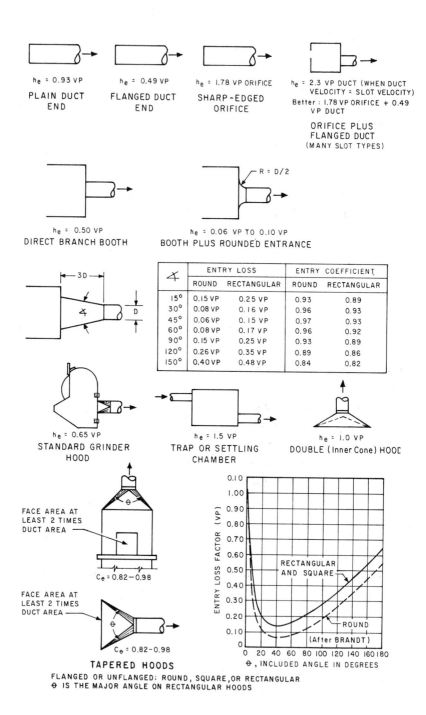

$h_e = 0.93\,VP$
**PLAIN DUCT
END**

$h_e = 0.49\,VP$
**FLANGED DUCT
END**

$h_e = 1.78\,VP\ ORIFICE$
**SHARP-EDGED
ORIFICE**

$h_e = 2.3\,VP\ DUCT$ (WHEN DUCT
VELOCITY = SLOT VELOCITY)
Better : 1.78 VP ORIFICE + 0.49
VP DUCT

**ORIFICE PLUS
FLANGED DUCT**
(MANY SLOT TYPES)

$h_e = 0.50\,VP$
DIRECT BRANCH BOOTH

R = D/2

$h_e = 0.06\,VP\ TO\ 0.10\,VP$
BOOTH PLUS ROUNDED ENTRANCE

∢	ENTRY LOSS		ENTRY COEFFICIENT	
	ROUND	RECTANGULAR	ROUND	RECTANGULAR
15°	0.15 VP	0.25 VP	0.93	0.89
30°	0.08 VP	0.16 VP	0.96	0.93
45°	0.06 VP	0.15 VP	0.97	0.93
60°	0.08 VP	0.17 VP	0.96	0.92
90°	0.15 VP	0.25 VP	0.93	0.89
120°	0.26 VP	0.35 VP	0.89	0.86
150°	0.40 VP	0.48 VP	0.84	0.82

$h_e = 0.65\,VP$
**STANDARD GRINDER
HOOD**

$h_e = 1.5\,VP$
**TRAP OR SETTLING
CHAMBER**

$h_e = 1.0\,VP$
DOUBLE (Inner Cone) HOOD

FACE AREA AT
LEAST 2 TIMES
DUCT AREA

$C_e = 0.82 - 0.98$

FACE AREA AT
LEAST 2 TIMES
DUCT AREA

$C_e = 0.82 - 0.98$

TAPERED HOODS
FLANGED OR UNFLANGED: ROUND, SQUARE, OR RECTANGULAR
θ IS THE MAJOR ANGLE ON RECTANGULAR HOODS

RECTANGULAR
AND SQUARE

ROUND

(After BRANDT)

θ , INCLUDED ANGLE IN DEGREES

ENTRY LOSS FACTOR (VP)

FIG. 13-2. Hood entry losses. From Industrial Ventilation,
10th ed., American Conference of Governmental Industrial Hygienists, Lansing, Michigan, 1968.

TABLE 13-1

Minimum Transport Velocities[a]

Material	Velocity (fpm)
Vapors, gases, smokes, fumes and very light dusts (small particles)	2000
Medium density, dry dusts	3000
Average industrial dust	4000
Heavy dusts	5000
Large particles	>5000

[a]From A. D. Brandt, Industrial Health Engineering, John Wiley, New York, 1947.

where L = length of duct to cause a loss of 1 VP (ft), C = roughness coefficient = 30 for old rough duct and 60 for new smooth duct = 55 for good design, D = duct diameter (in.), and Q = air flow (cfm). For rectangular duct sections, an "equivalent diameter" is often calculated as $\sqrt{(ab)}$ or as $2\,ab/(a + b)$ where a and b are the side dimensions of the rectangle. In terms of the actual head loss in conventional units,

$$h = 0.058\, LQ^{1.91}/(CD^5),\qquad\qquad (13\text{-}8)$$

where h = head loss (in. w.g.), L = length of duct (ft), and other symbols as above. Some designers prefer to use a head loss proportional to Q^2 rather than $Q^{1.91}$. The coefficients are empirical and for agreement should reflect a considerable difference (>2X for $Q = 10^4$ cfm) for the two powers. Figure 13-3 shows frequently used design relationships of the parameters V, Q, h, and D.

The various losses for duct fittings and connections are calculated in terms of the straight duct equivalent length by

$$L = aD^{1.183},\qquad\qquad (13\text{-}9)$$

where L = equivalent length of duct (ft), D = duct diameter (in.), and a = 1.338 for 90° elbows of radius 1.5 D; 0.87, radius 2.0 D;

0.73, radius 2.5 D (use 2/3 L for 60° elbows and 1/2 L for 45° elbows) = 0.577 for 30° entries; 0.87 for 45° entries. The head loss for a branch of a duct system is then determined in velocity heads; it is made up of 1 VP for starting the air movement plus the entry loss and the duct frictional losses.

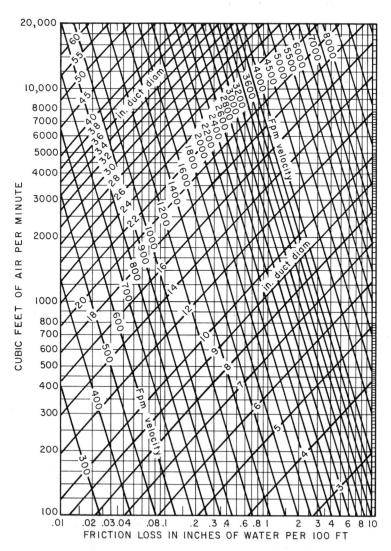

FIG. 13-3. Duct losses. Reprinted by permission of the American Society of Heating, Refrigerating and Air-Conditioning Engineers.

The design of a duct size for a single-branch exhaust system simply uses a diameter which will handle the volume required
by the hood at a velocity greater than the minimum transport velocity but smaller than that which would give an intolerable head
loss or noise. For multiple-branch duct systems, the head loss
in two or more branches up to a junction of the branches will
necessarily be equal; therefore, design of a branched system is
concerned with obtaining the design flows in each branch while
balancing the head losses.

The balance in the head losses in a branched system may
be obtained by either of two methods, namely, balanced duct
design or blast gate design. In balanced duct design the dominant
design parameter is the head loss in each of the branches. The
diameters of the ducts are set so that the required flows are met
with equal head losses in branches from a junction. In blast
gate design, the ducts are sized to provide the minimum transport velocities at tolerable head losses and then the lower head
loss branches are brought up to the higher ones through the use
of blast gates or dampers in the ducts.

EXAMPLE 13-1: Design a balanced-duct system for branches A
    and B of the layout shown. Minimum transport velocity
    = 4000 fpm.

a, 45° entry; b, 45° entry; c, 90° elbow (2 D); CL, cleaner
(8 in. w.g.); F, fan; 3, exit bonnet; 1 & 2, entry hoods

Branch A:

Max. diameter: $800/(\pi D^2/4) = 4000$ $D = 0.504$ ft = 6+ in.

VP for 6 in. D: $V = 4075$ $VP = (4075/4002)^2 = 1.04$ in. w.g.

$h_e = (1 - C_e^2)/C_e^2 = (1 - 0.55^2)/0.55^2 = 2.31$ VP

Equivalent length a and b: $L = 0.87D^{1.183}$ (Eq. 13-9)

$L = 0.87(6)^{1.183} = 0.87(8.32) = 7.24$ ft

Total duct length: $10 + 15 + 2(7.24) = 39.5$ ft

Total head: $1.00 + 2.31 + 39.5/[0.0365(55)(6)(800^{0.09})]$

$= 5.11$ VP $= 5.30$ in. w.g.

Branch B:

Max. diameter: $600/(\pi D^2/f) = 4000$ $D = 0.437$ ft $= 5.24$ in.

VP for 4 in. D: $V = 600/0.0873 = 6880$ fpm

VP $= (6880/4002)^2 = 2.95$ in. w.g.

$h_e = (1 - 0.82^2)/0.82^2 = 0.49$ VP

Equivalent length b: No change in straight-through
                    enlarging portion.

Total duct length: 10 ft

Total head: $1.00 + 0.49 + 10/[0.0363(55)(4)(600^{0.09})]$

$= 2.19$ VP $= 6.46$ in. w.g.

Balancing:

If 4 1/4-in. duct is available, it would give a total head
of 4.98 in. w.g. at a flow rate of 600 cfm or 5.30
in. w.g. at nearly 625 cfm.

Flow in Branch A could be adjusted to 900 cfm or smaller
size of duct selected for a total head of 6.47 in. w.g.

[Note: Most designers who are responsible for local exhaust
design use a columnar form for calculations (see
Industrial Ventilation).]

C. Fans

Fan selection is highly important to successful air pollution
control. Changes in the gas flow rate nearly always affect the
efficiency of the air-cleaning equipment and/or the process being
vented. The cost of air pumpage is often the dominant cost in air
pollution control, but adequate pumping capacity usually pays
dividends.

Straight- and backward curved-blade centrifugal fans are
most commonly used in air pollution control. The forward curved-
blade fans are less noisy, but they do not deliver sufficient head
for most applications. It is usually recommended practice to put
the fan after the cleaner so that the air moved is clean air; this

is especially desirable for particulate pollutants. An added advantage of such an arrangement is that leakage is into the dirty
air part of the duct rather than out of it. The straight-blade fan
is better for operating in dusty gas streams than the curved blade;
it has fewer blades and less tendency for buildup on the blades.
Other air movers which may be used include vane axial fans and
lobe pumps.

The considerations involved in the selection of a fan are
the volumetric flow rate (Q, cfm), static pressure across fan
(in. w.g.), temperature and density of gas stream, mechanical
and overall efficiency, noise produced, and special application
conditions. The latter include available space, power (ac, dc,
or turbine), material handled (dust, corrosive, explosive, sticky
particles), intake and discharge directions, and flexibility of
changing air flow (direct drive or belt drive, vari-speed motors).
The shape of the characteristic curves will aid the selection of
the fan type (see Fig. 13-4).

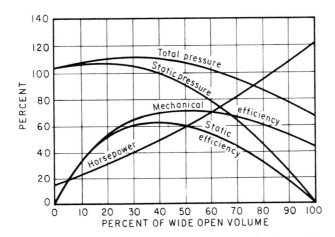

FIG. 13-4. Typical fan characteristics. Courtesy of Air
Moving and Conditioning Association, Inc., 1963.

Fan manufacturers furnish multirating tables of design data for homologous series of fans. Actual fan selection is made from the middle portion of the table where the efficiency is highest. For other than standard conditions, the actual flow rate is used with the actual static pressure to get the fan speed, then the horsepower is adjusted by a multiplying ratio of the density of the gas stream to the density of air at NTP (0.075 lb/ft^3).

When selecting fans, it is good practice to select a fan that is somewhat oversized for the anticipated application in order to allow for future changes. It is sometimes desirable to increase the flow rate of the exhaust system after the unit is in operation. This is often done by increasing the fan speed (S), where S may represent either the RPM or the tip speed of the fan blades. The fan laws will govern the relative changes according to

$$Q = aD^3S; \quad \text{s.p.} = bD^2S^2\rho_g; \quad \text{and} \quad HP = cD^5S^3\rho_g, \quad (13\text{-}10)$$

where Q = volumetric flow rate (acfm), D = diameter of fan (ft), S = fan speed (fpm), s.p. = static pressure (in. w.g.), ρ_g = density of gas stream (lb/ft^3), and HP = horsepower of fan (BHP). It should be noted that changing the fan speed (S) while holding D constant will change the flow rate directly with the speed (Q = fS), the static pressure with the square of the speed (s.p. $= gS^2$), and the power requirements with the cube of the speed (HP = hS^3).

III. CONDITIONING GAS STREAM

Conditioning the gas stream carrying the pollutants from a process may be desirable in order to facilitate cleaning or dispersion or to promote safety. Such conditioning is most often designed to change the temperature of the gas stream, usually cooling a hot stream but often reheating a cold gas stream. Materials may be added to the gas stream to condition the entrained

particles for collection, especially for changing the resistivity
for electrostatic precipitation. Precleaning may be used to remove
some constituent before the principal treatment is undertaken. A
waste gas stream may be diluted to prevent explosion and fire
hazards.

Cooling the hot gas streams from smelters, furnaces, and
hot processes may be done by: (1) diluting with ambient or other
cooler air; (2) radiation-convection from the carrying ducts and
pipes; (3) quenching with a water spray; (4) heat recovery with
a heat exchanger; and (5) a combination of these means.

Dilution is not economical for making large temperature
adjustments on large flows. It is a desirable method for very
small flows and for the final adjustment of temperature over a
small range after some other method has brought the temperature
very near that desired. Dilution increases the volume of gas to
be treated and the power for pumping. Dilution is accomplished
in most applications by the use of a branched-duct opening with
a temperature-controlled valve or damper that admits the amount
of dilution air necessary. This arrangement serves as a safety
guard even on systems where the valve is ordinarily fully closed,
especially before bag filters. Calculations are made for dilution
cooling by a heat balance using the enthalpies of the gas stream,
the dilution air, and the final mixture. The specific heat of air
is about 0.24 Btu/lb-$^{\circ}$F and that of water vapor about 0.48 Btu
/lb-$^{\circ}$F.

Radiation-convection will take place to some extent through
the normal ductwork and piping. Design for this type of cooling
is calculated on the basis of furnishing sufficient exchange area
between the pipes and the atmosphere to allow the necessary
heat exchange according to

$$Q_h = UA \, \Delta t_m, \qquad\qquad (13\text{-}11)$$

where Q_h = heat transfer rate (Btu/hr), U = heat transfer coefficient (Btu/hr-ft^2-^{O}F), A = area of heat transfer (ft^2), Δt_m = log-
mean temperature difference (^{O}F) or

$$\Delta t_m = \frac{(t_i - t_a) - (t_o - t_a)}{\ln[(t_i - t_a)/(t_o - t_a)]} \,,$$

t_k = temperature (^{O}F): k = i, inlet = o, outlet = a, ambient.
Radiation-convection cooling units are usually standing loops
of black iron pipe that are 1 to 2 ft in diameter. Values of U for
Eq. 13-11 are calculated from theoretical and empirical concepts
of heat transfer or are obtained from tables or nomographs in hand-
books. As a rough estimate, U may be taken for normal conditions
as 2 Btu/hr-ft^2-^{O}F (usual range 1.5 to 3.0). The radiation-
convection unit may be designed such that some or all of the
unit is bypassed if the desired temperature is reached before the
air has passed through all the loops. Some cooling units have
simply split the gas stream and passed part of the flow through
the unit as necessary to reach the desired temperature when the
streams were recombined. Such a procedure results in lowering
the temperature of the cooled portion below the dew point and
causes corrosion of the tubes. Radiation-convection coolers
are normally applied ahead of bag filters that clean the gas
streams from reverberatory and blast furnaces and other hot processes such as kilns and cupolas and they are usually applied
after quenching.

Quenching a hot gas stream with a water spray is most
often used because of its effectiveness, simplicity, and low cost.
The water vapor adds to the volume of gas, but as a result of the
high heat of vaporization for water (970 Btu/lb), the added volume
is relatively small.

EXAMPLE 13-2: A flue gas stream from a reverberatory furnace
at a lead smelter has a flow rate (Q) of 20,000 acfm at
1800°F. It contains 15%$_v$ of water vapor. The gas is
cooled to 275°F for filtration by a combination of water
quench (1800° to 800°F) and radiation-convection in pipe-
work (800° to 275°F). Water temperature is 62°F.
(a) What is the final volume of gas to be filtered?
(b) What is the moisture content of the gas to the filter?

Btu removed by quench:

$$0.85 \times 20,000 \text{ acfm} \times 0.075\left(\frac{68 + 460}{1800 + 460}\right)\frac{\text{lb}}{\text{ft}^3} \times 0.24 \text{ Btu/lb-}^\circ\text{F}$$

$$\times (1800^\circ - 800^\circ\text{F}) + 0.15 \times 20,000 \text{ acfm} \times 0.0464$$

$$\left(\frac{68 + 460}{1800 + 460}\right)\frac{\text{lb}}{\text{ft}^3} \times 0.48 \text{ Btu/lb-}^\circ\text{F} \times (1800^\circ - 800^\circ\text{F})$$

$$= 71,500 + 15,600 \text{ Btu/min} = 87,100 \text{ Btu/min}$$

Water required:

Latent heat	970 Btu/lb	
Sensible heat	432 Btu/lb	$= (800 - 212)0.48 + (212 - 62)1.0$
Total	1402 Btu/lb	

$$\frac{87,100 \text{ Btu/min}}{1402 \text{ Btu/lb}} = 62.1 \text{ lb/min}$$

Volume at 800°F:

$$20,000\left(\frac{800 + 460}{1800 + 460}\right) \text{ cfm} + 62.1 \text{ lb/min}\left(\frac{385 \text{ ft}^3}{18 \text{ lb}}\ \frac{800 + 460}{68 + 460}\right)$$

$$= 11,150 + 3170 \text{ cfm} = 14,320 \text{ cfm}$$

% Moisture:

$$\frac{[3170 + 0.15(11,150)]100}{14,320} = 33.8\%_v$$

Volume at 275°F:

$$14,320\ \frac{275 + 460}{800 + 460} = 8350 \text{ cfm}$$

The obvious solution to cooling a gas stream is the installation of a heat recovery boiler unit; however, because of the dirtiness of the gas streams and the problems from such dirtiness, this method has lagged. It should increase in popularity with the rising cost of fuel. One type of air cleaner currently under development uses hot water and steam generated by recovery boilers to scrub particulates from the gas stream (see Chapter 21).

In many applications quenching is used to reduce the gas temperature to about 800°F, then radiation-convection cooling is applied to lower the temperature to 250° to 450°F.

Gas reheating may be applied to prevent the formation of visible fog in the plume or for lifting the plume to aid in dispersion or for carrying the fog over a nearby street or roadway.

When a reheater is used, it is normally placed downstream of
the gas cleaner, which is usually a scrubber. Heaters have been
used on cooling tower effluents in order to prevent fog problems;
it is not the usual function of the heaters to prevent the formation
of the fog but rather to lift the fog to permit wind carriage across
the street or roadway and thereby avoid reduced visibility.

Various materials have been added to gas streams to lower
the resistivity of the particles for electrostatic collection. These
include water, sulfuric acid, ammonia, and sulfur trioxide. This
practice will be discussed further in Chapter 22. Limestone and
dolomite which have been added to the combustion chamber to
remove sulfur dioxide (see Chapter 24) have resulted in raising
the resistivity and lowering the efficiency of electrostatic precipitation of particles.

Precleaning is seldom applied as a roughing process because
it has not been found economical. It is possible to put air cleaners in series and the order of the cleaning may be a significant
factor. For example, if particulates are to be removed and a
gaseous constituent is to be reacted catalytically, the particulate
removal should take place ahead of the catalyst to prevent fouling
or poisoning of the catalyst. Similarly, the particulates should
be removed before adsorbing beds. In the past, roughing filters
have been placed ahead of absolute filters; however, the practice
has often been abandoned as not being worthwhile in cases where
the particle loadings were not too heavy. Settling chambers,
cyclones, and louver or baffle cleaners have been used as precleaners, but their use is now more infrequent.

Particles may be built up in size for improving collection
by agglomeration, condensation on the particles, or impaction
with water droplets. These methods are described later.

Preventing fire and/or explosion in gas streams may be
effected by keeping the material above the upper explosive limit

(UEL) or below the lower explosive limit (LEL), usually the latter, or by limiting the flame propagation in the system. The guidelines for these practices are discussed in Chapter 18.

IV. HANDLING AND DISPOSAL OF COLLECTED MATERIAL

Collected pollutants may be solid, liquid, or gaseous in form. Sometimes the form desired is the determining factor in the selection of an air-cleaning method. The ultimate disposal of a collected material depends on the material. The usual alternates are the following: (1) If the material is the product or a valuable by-product (e.g., as in carbon black collection, cement from cooling and grinding the clinkers, vapors escaping from the storage of oil products), it may be marketable as collected or with simple treatment; (2) If the material is raw material that has escaped the processing, it may simply be routed back to the process with or without treatment (e.g., kiln dust in cement or lime manufacturing); (3) If the material is innocuous, it may be put into the environment in such a way that it does not cause a pollution problem, usually after some condtioning (e.g., the limestone dust from quarrying and crushing operations); and (4) If the waste is objectionable, as it often is, it may require extensive treatment before ultimate disposal to the environment (e.g., strongly acid or alkaline materials, toxic materials, radioactive wastes).

Solid materials may be collected dry by electrostatic precipitation, bag filters, and mechanical or inertial collectors. If the material is to be put into the environment, it will usually require care and often treatment to prevent resuspension of dusts; the treatment is often just wetting with water. Where high heat is available, the solid particles can be melted into a slag which is much easier to handle than the bulk particles. The solids are usually placed in a land fill.

The solid polluting materials may be collected in suspension with water or put into such a suspension after collection. Suspended solids may result from precipitation of dissolved materials. Water streams with suspended solids are treated by sedimentation and/or filtration that may or may not be preceded by chemical coagulation. The treated liquid stream may be recirculated as scrubbing or transporting liquor or it may be put into a waterway; however, the latter alternative is rapidly disappearing with increasingly strict water pollution regulations.

Gaseous pollutants which are absorbed in scrubbing liquids or adsorbed on carbon are removed as concentrated gas streams. The concentrated gas may be valuable enough to market, but often it will have to be burned or chemically neutralized. The gas stream may have heating values sufficient to support combustion or the small volume may permit economic addition of fuel for combustion.

Radioactive wastes are treated much the same as other materials, except that very long-term storage of the wastes (up to 600 or 800 years) puts a high premium on concentrating the wastes into small volumes. The gas streams with short-lived radioactive components may simply be stored without treatment until decay takes place.

Pollutants which are sticky, explosive, abrasive, or corrosive require special precautions in handling. These types of materials often dictate the treatment and ultimate disposal as well as the method of collection of the pollutant.

V. EQUIPMENT FOR PERIPHERAL OPERATIONS

Much of the equipment that is vital to the success of air pollution control is ancillary to the actual collection equipment. This includes the exhaust equipment previously described, deflecting vanes, collecting hoppers, hopper valves, conveying equipment, treatment equipment, process monitoring devices, sampling and access ports, ladders and platforms for access, insulation (thermal and acoustical), structural supports and foundations, and pumps and hydraulic equipment.

Deflecting vanes are used to distribute the air flow for reducing head losses or improving the efficiency of collection. Deflecting vanes often make a large difference in both the cost and success of an air pollution control process. They may result in savings if they are used properly or losses if they are misused. The most frequent abuse of deflecting vanes lies in making the transition section where the vanes are employed too short for the vanes to accomplish their purpose.

Hoppers are placed on the bottoms of particulate collectors to hold the materials collected and to funnel them into the removal and transporting mechanisms. They are usually pyramidal or conical in shape with sides steep enough for the collected materials to slide down. Bulking of collected solids often causes a problem in removal of the hopper material, and schemes have been devised to alleviate bulking, bridging, and sticking, including rappers, vibrators, and rubber hoppers that are periodically struck with sledge hammers. The hopper valves are designed to remove the material from the hopper without dustiness or leakage of air despite a pressure differential between the hopper and the receiving device (see Fig. 13-5).

Conveying devices for collected materials must be designed to be compatible with the materials handled. Screw conveyors may be used with some solids, but not with sticky substances. The particles may be put into a slurry for pumping. Liquids and gases can be piped to the treatment or disposal site.

Sampling and access ports should be located on stacks and/or ducts for convenience in sampling. They must be large enough to accommodate appropriate equipment. Regulations often specify the locations, sizes, and accessibility of these ports. A sampling platform must be provided for the port.

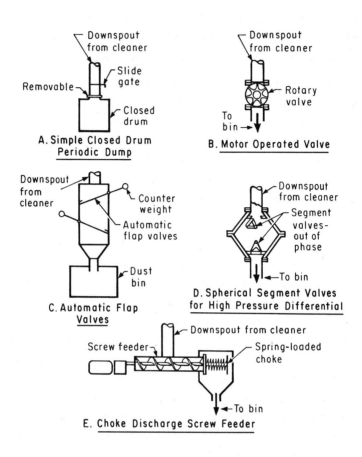

FIG. 13-5. Air cleaner hopper valves. From "Cyclone
Dust Collectors," American Petroleum Institute, New York.

VI. SUMMARY

Auxiliary operations often make the difference between
satisfactory and unsatisfactory air pollution control. Polluting
emissions must be captured, contained, and collected in air-
cleaning methods. These procedures are carried out by properly

designed and operated hoods, ducts, and collectors. Necessary adjuncts to the successful cleaning process are the peripheral devices that make the process work or the operation easier. The collected material must be disposed of without creating a new pollution problem. The importance of the procedures and operations listed in this chapter to the overall air-cleaning method can hardly be overemphasized.

PROBLEMS

1. What minimum transport velocity should be used for a dust with particle sizes to 400-μm diameter and density of 5.2 gm/cm^3?

2. It is desired to collect the exhaust gases at a customs inspection station with slots in the curbs of the 7 ft-wide lanes. The slots should be 4 ft long by 3 in. high and should be flush with the surface of the lane (drainage to center). What air volume needs to be pulled through the slot to give 100-fpm capture velocity halfway across the lane? Assume that the bottom of the slot is flanged.

3. Assume that Fig. 13-3 is a plot of Eq. 13-8. Use the 6000-cfm and 3000-cfm point information to solve for the coefficient (C) used in the plot.

4. Find parameters a, b, and c in $h = aQ^b/D^c$ for Fig. 13-3. Does your fitted equation agree with Eq. 13-8?

5. Complete the design of the layout in Example 13-1, including the horsepower of the fan with a 60% efficiency.

6. What is the area of radiation cooler required for Example 13-2 if the heat transfer coefficient is 2.15 Btu/ft^2-hr-$^\circ$F and the ambient temperature reaches 100°F?

BIBLIOGRAPHY

Air Pollution Manual, Part II: Control Equipment, American Industrial Hygiene Association, Detroit, Michigan, 1968.

Brandt, A. D., Industrial Health Engineering, John Wiley, New York, 1947.

Constance, J. D., "Estimating Exhaust-Air Requirements for Processes," Chem. Engr., 77:17, 116-118, August 10, 1970.

Danielson, J. A., Ed., Air Pollution Engineering Manual, U. S.
    Public Health Service Publ. No. 999-AP-40, National
    Center for Air Pollution Control, Cincinnati, Ohio, 1967.

Hemeon, W. C. L., Plant and Process Ventilation, 2nd. ed.,
    The Industrial Press, New York, 1963.

Industrial Ventilation, 10th ed., American Conference of Govern-
    mental Industrial Hygienists, Lansing, Michigan, 1968.

Stern, A. C., Ed., Air Pollution, Vol. III: Sources of Air Pollu-
    tion and Their Control, Academic Press, New York, 1968.

Section II: Prevention

Prevention of air pollution problems is our goal. With practically every air pollution problem, the high concentration near the point of emission is the trouble factor; therefore, any procedure that will lower the concentration to an innocuous level should be considered successful control. In some cases the problem may be alleviated by a process change and, in others, by dispersion into the environment; both of these control methods can often be adequately applied without resorting to more expensive air cleaning.

Chapter 14

PREVENTING FORMATION/RELEASE OF POLLUTANT

I. INTRODUCTION

It is often possible to achieve air pollution abatement by
preventing the formation and/or release of the pollutant. Controls in this category range from simple housekeeping to total
changeovers of the process. The costs of control by these means
may be more economical than air cleaning and certainly merit
careful consideration.

II. PROCESS SUBSTITUTION

There are many instances in which operational changes can
help in air pollution control, such as changing the process, the
fuel, or the process materials.

A. Changing the Process

Process curtailment, even to shutting down, during adverse
meteorological conditions continues to be an important method of
preventing air pollution problems. ASARCO has been using a
"Nelson closed loop" system of control at El Paso, Tacoma, and
other locations for several years. Sulfur dioxide sampling stations surround the plant; the sampled concentrations are tele-
metered to the plant where meteorologists use the data for computer input to forecast possible problem ground-level concentrations; the meteorologist shuts down as much of the sulfur dioxide-
emitting process as required to meet the regulations and to prevent
problems. Other smelters in Arizona and Canada are adopting this
system of problem abatement. TVA has announced that it intends

to cut back on steam-electric generation with high sulfur coal
during periods of atmospheric stagnation.

Good housekeeping can make a strong contribution to air
pollution abatement. For example, process spills should be
cleaned up immediately to prevent the material from becoming
airborne as particles or vapors. First (see Bibliography) details
the importance of such actions when the materials are putrescible.

Within the last 40 years or so, there has been a halving of
fuel required to generate a kilowatt of electricity. However,
much of the real contribution of this achievement to air pollution
abatement has been lost through two factors, the increased use
of cheaper electricity and the release of tremendous quantities
of pollutants at a single location instead of the smaller amounts
at several locations (a large fraction of the increased efficiency
is made possible by the increased size of the plant). Improved
insulation of homes can result in substantial reductions in the
amount of fuel used for heating and cooling.

Many transportation abatement schemes involve process
changes. These include external combustion engines for the
present internal combustion engines, electric cars, mass transit,
or even bicycles. Mass transit may well be the answer to auto
exhaust pollution in large metropolitan areas. The downtown
areas could be blocked from all vehicular traffic except trackless
electric trolleys.

The smelting industry and the papermaking industry are
trying to rid themselves of their notorious reputations for emitting
highly objectionable sulfurous materials by major process changes--
hydrometallurgical separations of ores and the manufacture of
paper without the use of sulfides.

Agriculture has lowered its contribution to air pollution by
plowing practices that reduce wind erosion, better methods of
crop spraying or dusting to prevent wind drift, and burning field

stubble and other materials only during favorable meteorological
conditions. There is still room for much improvement in the
agricultural and food industry. Cattle feed lots and large swine
raising operations are having troubles with their neighbors because
of their emissions, especially odors, dusts, noise, and flies.

B. Changing Fuels

In England, from the 13th century until the present, wood
has been used instead of coal as a smoke abatement procedure.
In the United States, a marked emphasis has been placed on the
use of hard (anthracite) coal instead of soft (bituminous) coal
for abating smoke.

The numerous sulfur regulations, both on fuel sulfur content and on stack emissions of sulfur compounds, have resulted
in premium prices for low sulfur fuels, feverish activity in the
development of practical methods of sulfur removal from the fuel
and the stack, and the search for alternate energy sources that
are economically feasible. Because of its low pollution potential,
natural gas is in great demand and it is being shipped very long
distances by pipelines and by ships (in liquefied form). The
most promising way of using the large coal reserves with minimum
air pollution appears to be coal gasification. The sulfur and some
other unwanted materials can be removed from gas much more
readily than from the solid coal or from the stack gas.

Nuclear fuel can be used in some current design power
plants with practically no release of radionuclides to the environment. Although the risks from such plants are extremely low,
much of the public still views nuclear power with alarm; therefore,
nuclear-fueled plants are not filling the role they should have
occupied while the problem of sulfur removal from coal plants is
being solved.

Because of the current concern for heavy metals in the
environment, there will probably be regulations against many of
the heavy metals in fuels. The regulation against adding lead
to gasoline has already been scheduled for phasing in to start in
1974. There are other heavy metals which occur naturally in
fuels or get there through current processing techniques. Most
interest is now centered on mercury and vanadium, but it will
surely expand to other heavy metals.

C. Changing Materials

Of all the air pollution abatement changes that have been
made in materials, the most sweeping are probably those in the
paint industry. Regulations against the release of solvents from
applied paint have led to the development of water-based paints,
various plastic and resin coatings, and now dry painting efforts.
The paint industry has had to make these changes while changing
the pigments to meet the lead limits against ingestion poisoning.

Rose et al. cited the foundry application of a flux to the
surface of molten brass in order to reduce the evaporation and
the resulting fume emissions. Lead smelters have also used a
similar flux on the crucibles for transporting the molten lead and
on the surfaces where the liquid lead is poured. Other fluxing
operations have gone through material substitutions, including
a bauxite flux for fluorspar at open hearths to prevent fluoride
emissions and borate salts for elemental sulfur in magnesium
casting.

Sometimes the form of a material can significantly alter
the emissions during its use. In the iron and steel industry,
the use of sintered ores has resulted in less dust emissions.
Dust emissions from rock crushing can be lessened by the simple
expediency of feeding wetted rock to the crusher.

III. ENCLOSING PROCESS

Storage facilities may be enclosed to prevent losses to the atmosphere. The oil industry has covered many storage tanks and other evaporative sources of hydrocarbons and put in vapor recovery units on these and on other operations, including the filling and emptying procedures for tanks. Vapor recovery units have even been put on automobile gas tanks and carburetors.

Chemical process plants often have excessive leakage around pump and valve packings, pipe cracks, and other locations. Many such leaks can and should be stopped by instituting better maintenance procedures.

Recent experiences with uranium mill tailings have caused an increased awareness of the need for emission controls on tailings piles and on stockpiles of ores and ore concentrates. Wind erosion of ore concentrate piles at a lead smelter has been controlled by putting a crust over the piles with an emulsion spray. The spray has also been used to crust materials being shipped in open hopper or gondola cars. The alternative to stabilizing the piles against wind erosion would appear to be the use of covered bins for storing the materials.

IV. RELOCATION OF PROCESS

The coal-burning power plant is probably the best example of a process for which complete control is infeasible, and, therefore, the plant should be so located that a minimum population exposure results. The development of ultra-high voltage transmission in the past few decades has made it possible to locate power plants many hundreds of miles from the population centers they serve.

V. SUMMARY

There are times when the best solution to air pollution abatement is to prevent the formation and/or release of the pol-

lutant. This is especially true for those processes which emit
substances that would be very expensive to remove with current
technology for air cleaning. Although many of the possible
abatement procedures require careful technical study to delineate
the problem and to initiate the solution, good housekeeping can
be done by all processes but is often ignored.

PROBLEMS

1. What are the relative advantages of hooding individual Soder-
   berg pots in an aluminum smelter versus using the whole
   building as a hood to collect the emissions for cleaning?

2. Answer Problem 1 for open hearth furnaces.

3. What are some processes other than power plants which might
   be relocated to prevent air pollution problems?

4. Copper smelting produces a blister copper which is then
   electrolytically refined. Look into the reasons for not
   electrolytically separating the copper from the high sulfur
   ore concentrate directly.

5. Compare ammonium nitrate blasting with TNT blasting in
   regard to the air pollution emissions.

BIBLIOGRAPHY

First, M. W., "Process and System Control," Air Pollution,
   Vol. III, 2nd ed., Chap. 41 (A. C. Stern, Ed.), Academic
   Press, New York, 1968.

Rose, A. H., Jr., D. G. Stephan, and R. L. Stenburg, Prevention
   and Control of Air Pollution by Process Changes or Equip-
   ment, SEC TR A58-11, Public Health Service, Cincinnati,
   Ohio, 1958. See same title in Air Pollution, World Health
   Organization: Monograph Series, No. 46, pp. 397-343,
   Geneva, 1961.

Chapter 15

TALL STACKS--DISPERSION FOR DILUTION AND DECAY

I. INTRODUCTION

Tall stacks have occupied an important place in the prevention of air pollution problems. Stack heights are increasing
because of the increasing size of the sources emitted through
the stacks and the stiffening of air pollution regulations. The
use of tall stacks for the dispersion of sulfur dioxide will probably
continue until coal gasification for power plants and chemical
separation of pyritic ores for smelters have developed to the
practical stage. Although tall stacks have been used for gases
other than sulfur dioxide, particularly for radioactive gases, and
for small particles, their principal applications have been and
continue to be for dispersion of sulfur dioxide.

II. THEORY

The purpose served by tall stacks is the dilution of the
plume concentrations to acceptable levels before the material
reaches the ground. This purpose is accomplished by injection
of the plume into the atmosphere at a height which gives eddy
dispersion time to carry out the dilution. Tall stack dispersion
certainly prevents local problems and it may help regional problems caused by additive sources as well. At least there is some
evidence that discharging the sulfur dioxide high into the air
may shorten its lifetime in the atmosphere.

Tall stacks serve a useful purpose not only for the normal
meteorological conditions that are ordinarily used in design but
also for the unusual stagnating conditions which result in the

79

high ground concentrations (see Table 15-1). Tall stacks often inject the pollutant into the atmosphere above the nocturnal inversion, thus preventing much of the fumigation problem caused when the inversion breaks up from daytime heating and the pollutant collected just below the inversion is carried to the ground level. The tall stack is also used to put the discharge from a plant in a valley location above the walls of the valley where there are more winds and less chance for stagnation buildup.

Specific air pollutants are removed from the atmosphere by deposition on the surfaces of airborne particles and/or the ground interface or by chemical change to other materials. The rates of removal of pollutants are usually referred to as being first order and characterized by half-lives; however, the lifetimes of atmospheric pollutants are not well known, not even those for the common pollutants.

TABLE 15-1

Criteria for National Air Pollution Potential Advisory

A. Stagnation area
  1. wind speed at 5000 ft above station $u \leq 10$ m/sec
  2. temperature change at 5000 ft during
     last 12 hr $\Delta T \geq -5^{\circ}C$
  3. 500 mb vorticity[a] $w \leq 100(10^{-6})$ sec^{-1}
  4. 500 mb vorticity change last 12 hr $\Delta w \leq +30(10^{-6})$ sec^{-1}
  5. observed precipitation or   pptn ≤ 0.01 in.
     PE relative humidity   RH $\leq 80\%$

B. Mixing and ventilation
  1. morning mixing height   MMH ≤ 500 m
  2. morning wind $u \leq 4$ m/sec
  3. afternoon ventilation[b] $V \leq 6000$ m^2/sec
  4. afternoon transport wind $u \leq 4$ m/sec

C. Prevailing area   (4° lat X long) $A \geq 58,000$ naut mi^2

D. Duration alert must continue $t \geq 36$ hr

E. May add to area

---

[a]$w = 2$ X angular velocity in radians/sec.

[b]Ventilation = mixing height X wind speed.

III. DESIGN

The tall stack must be designed for its primary function--
the reduction of the ground level concentrations of the pollutant,
for its compatibility with the process being vented (especially
with respect to pressures and volumes or velocities), and for its
structural and esthetic integrity.

A. Functional Design

The functional design of tall stacks must set the effective
stack height to meet the ground level concentration allowed by
the regulations or a consideration of effects. All the parameters
which enter into the calculations for obtaining the required effective stack height relate to process design.

At some point downwind from the tall stack, the plume will
spread by eddy dispersion until the gases and/or the particles
from the plume reach the ground. There are many formulas for
calculating the ground level concentration from a continuous
point source such as a stack. The best one in current use is
probably that of Pasquill and Gifford:

$$X = \frac{Q\,10^6}{\pi \sigma_y \sigma_z u}\ \exp\left[-\frac{1}{2}\left(\frac{y^2}{\sigma_y^2} + \frac{H_e^2}{\sigma_z^2}\right)\right] , \qquad (15-1)$$

where X = ground concentration (ppm_v or ug/m^3), Q = emission
rate of pollutant (m^3/sec or gm/sec), y = horizontal distance
from the centerline of the plume normal to the x (downwind)
direction (m), u = wind speed (m/sec), σ_y and σ_z = dispersion
parameters in y and z (vertical) directions, respectively (m), and
H_e = effective stack height or height of plume (m)(see Fig. 15-1
and Table 15-2). A more detailed discussion of this and other
formulas relating to dispersion can be found in Chapter 3, Atmospheric Transport of Pollutants. The X calculated by Eq. 15-1 is
the maximum short-term concentration expected and needs to be
adjusted for extrapolation to any time period other than about 10
min (see Chapter 6).

TABLE 15-2

Stability Categories

Wind speed @ 10 m (m/sec)	Daytime Incoming solar radiation			Nighttime	
	Strong	Moderate	Slight	Thinly overcast or ≥ 4/8 low cloud	≤ 3/8 Cloud
<2	A	A-B	B	--	--
2-3	A-B	B	C	E	F
3-5	B	B-C	C	D^a	E
5-6	C	C-D	D	D	D
>6	C	D	D	D	D

^{a}The neutral class, D, should be assumed for overcast conditions during day or night. From D. B. Turner, Workbook of Atmospheric Dispersion Estimates, NCAPC, U. S. Public Health Service Publ. No. 999-AP-26, Cincinnati, Ohio, 1967.

EXAMPLE 15-1: What maximum ground level concentration of sulfur dioxide should be expected for an H_e of 400 m, type D stability, u of 3.8 m/sec, and Q of 20,000 lb/hr? Where does this maximum occur?

$Q = 20,000$ lb/hr $= 2520$ gm/sec $= 0.945$ m^3/sec

From Fig. 15-3: Max. at 35 km $Xu/Q = 2 (10^{-7})$ m^{-2}

$$X = \frac{2 (10^{-7})(2520)(10^6 \, \mu g/gm)}{3.8 \, m/sec} = 132.5 \, ug/m^3 = 0.05 \, ppm_v$$

[Note: Use of 0.945 m^3/sec instead of 2520 gm/sec gives the answer in ppm$_v$ directly.]

Decay of the pollutant is calculated using the equation

$$f = (1/2)^{t/T_{\frac{1}{2}}} = \exp(-0.693 \, t/T_{\frac{1}{2}}), \qquad (15-2)$$

where f = multiplying factor for X, t = time since emission (hr), and $T_{\frac{1}{2}}$ = half-life of pollutant (hr). The available sulfur dioxide decay data show that its half-life depends strongly on the relative humidity (RH). The half-life of sulfur dioxide may be as short as 15 min for 100% RH (fog); 1.2 hr for 80% RH; 2.5 hr for 50% RH; and even days for RH values less than 30%. For settling particles, $T_{\frac{1}{2}}$ may be estimated as $0.693 \, H_e/u_t$, where u_t is the terminal settling velocity of the particles (see Part A, Chapter 5, Properties

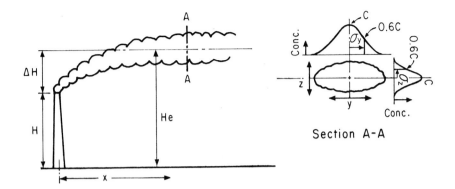

FIG. 15-1. Parameters for dispersion calculations.

of Gaseous and Particulate Matter). The half-lives of carbon monoxide and hydrogen sulfide may be on the order of 1.2 mo and 2 hr, respectively, but they vary with the height of release and the type of surroundings into which they are released.

Another method which is often used for calculating the decayed concentration X is the inclusion of a deposition velocity in the exponential term of Eq. 15-1. For settling particles, H_e may be replaced by H_e minus the settling distance of the particles. The settling distance may be shown as $x \tan(u_t/t)$ or as $u_t t$ or as $u_t x/u$. This procedure is equivalent to inclining the plume below the horizontal. For gases the deposition velocities have been reported in the range of 0.1-10 cm/sec with values of 0.1-2.8 cm/sec having been measured for iodine-131 in fission products clouds.

Although several investigators have used a power law relation between σ_y and σ_z and x (i.e., $\sigma_i = ax^b$), the plots in Fig. 15-2 are obviously not straight lines. As a result it is difficult to obtain the maximum X by Eq. 15-1. Figure 15-3 may be used to get the point of maximum X and its value, as was done in Example 15-1 above.

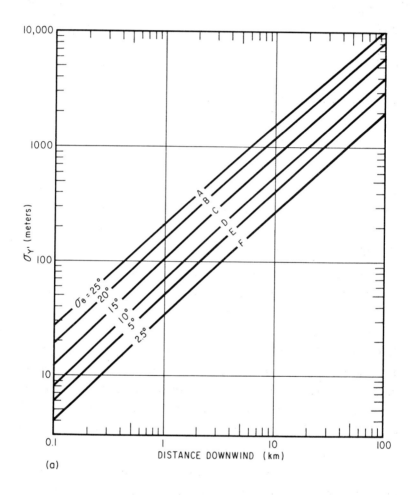

FIG. 15-2a. Values of σ_y for dispersion estimates. From
D. B. Turner, <u>Workbook of Atmospheric Dispersion Estimates</u>,
NCAPC, U. S. Public Health Service Publ. No. 999-AP-26,
Cincinnati, Ohio, 1967. σ_θ = standard deviation of wind azimuth
fluctuations.

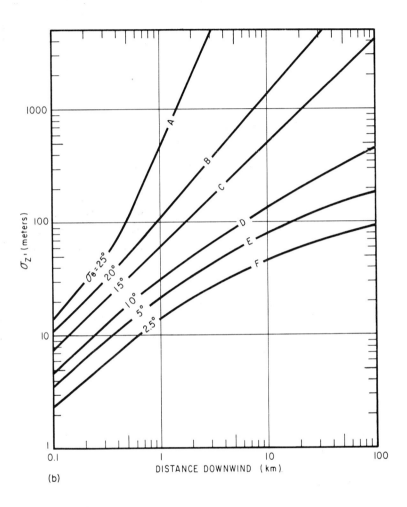

(b)

FIG. 15-2b. Values of σ_z for dispersion estimates. From
D. B. Turner, <u>Workbook of Atmospheric Dispersion Estimates</u>,
NCAPC, U. S. Public Health Service Publ. No. 999-AP-26,
Cincinnati, Ohio, 1967.

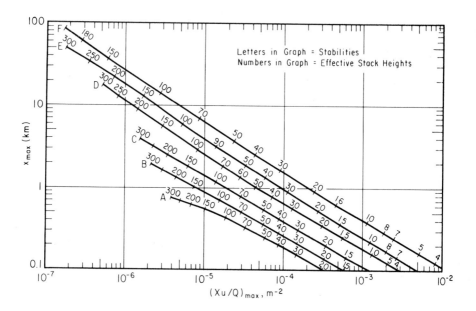

FIG. 15-3. Maximum ground concentrations and their
distances. Letters = stabilities; numbers = effective stack heights.
From D. B. Turner, <u>Workbook of Atmospheric Dispersion Estimates</u>,
NCAPC, U. S. Public Health Service Publ. No. 999-AP-26,
Cincinnati, Ohio, 1967.

It should be kept in mind that Eq. 15-1 is intended for the
ground concentration and has a factor of 2 included in its denominator to account for the assumed total reflection of the plume
from the ground.

B. Process Design

After H_e has been determined, the stack parameters are
set to meet the H_e in conformance with the process considerations
and the economics of meeting the H_e by the various combinations

of influencing factors. H_e is made up of two parts, the actual stack height (H) and the plume rise (ΔH). The most economical stack combines H and ΔH in optimum proportions, within practical constraints.

The factors which are needed for stack design are the gas flow and temperature with their ranges and fluctuations, dewpoint of gas, corrosives in gas, data on duct openings to stack (number, size, height above foundation), draft required on ducts, elevation of site, topographical map of surrounding buildings and terrain, and meteorological data (wind rose, temperatures, humidities, and other climatological information).

The plume rise is often calculated from a modified Holland equation such as

$$\Delta H = \frac{1.5\,vD + 4.09\,(10^{-5})\,Q_H}{u} , \qquad (15-3)$$

where ΔH = plume rise above the top of the stack (m), v = stack exit velocity (m/sec), D = exit diameter (m), Q_H = excess heat of stack gas (cal/sec) = $Q_m c_p (T_s - T_a) = 0.24\,Q_m(T_s - 20^{\circ}C)$, Q_m = mass emission rate of stack gas (gm/sec), T_s = temperature of stack gas ($^{\circ}C$), T_a = temperature of ambient air = $20^{\circ}C$ for normal conditions = $32^{\circ}C$ or even $40^{\circ}C$ for minimum plume rise in hot weather, and u = wind speed (m/sec). The ΔH calculated from Eq. 15-3 is for neutral atmospheric conditions; it should be adjusted upward about 20% for the unstable case and downward about 20% for the stable condition.

EXAMPLE 15-2: How much additional rise should be gained by a change of stack exit temperature from 150° to 160°C if the exit velocity is 90 ft/sec from a 24-ft diameter stack into a 3.8-m/sec wind? Assume neutral conditions.

$\Delta H = [1.5\,vD + 4.09\,(10^{-5})\,Q_H]/u$

$v = 27.5$ m/sec; $D = 7.32$ m; $T_a = 20^{\circ}C$

$Q_m = [(\pi/4)(24^2)(90)\ \text{ft}^3/\text{sec}][0.075\ (293/423)(454)\ \text{gm/sec}]$

$\qquad = 9.61\,(10^5)$ gm/sec

At 150°C: $Q_H = 9.61(10^5)(0.24)(423 - 293) = 3(10^7)$ cal/sec
At 160°C: $Q_H = 3.23(10^7)$ cal/sec
$\Delta H_{150} = [1.5(27.5)(7.32) + 4.09(10^{-5})(3 \times 10^7)]/3.8$
$= 403$ m
$\Delta H_{160} = [302(433/423) + 1230(140/130)] = 429$ m
Difference: $429 - 403$ m $= 26$ m

Plume rise formulas that incorporate a buoyancy flux parameter are increasing in popularity. The TVA (see Montgomery et al.) has made extensive plume rise observations and formula development. They now recommend the use of

$$\Delta H = 173 \, F^{1/3}/[u \exp(0.64 \Delta\theta/\Delta z)], \quad @ \ x = 1824 \text{ m} \quad (15-4)$$

where F = buoyancy flux parameter $(m^4/sec^3) = gvR^2(\rho_a - \rho_s)/\rho_a$, [g = acceleration of gravity (m/sec^2), R = inside radius of stack exit (m), ρ_a and ρ_s = densities of ambient air and stack gas (gm/m^3)], $\Delta\theta/\Delta z$ = potential temperature gradient (°C/100 m), and other terms are same as above. For downwind distances (x) less than the full plume rise distance, they recommend

$$\Delta H = ax^b F^{1/3}/u, \qquad\qquad (15-5)$$

where: with neutral conditions $(-0.17 < \Delta\theta/\Delta z \leq 0.16$, average 0.05) and $x \leq 3000$ m, a = 2.50 and b = 0.56; with moderately stable conditions $(0.16 < \Delta\theta/\Delta z \leq 0.70$, average 0.43) and $x \leq 2800$ m, a = 3.75 and b = 0.49; and with very stable conditions $(0.70 < \Delta\theta/\Delta z \leq 1.87$, average 1.06) and $x \leq 1960$ m, a = 13.8 and b = 0.26. The solution of Example 15-2 with Eq. 15-4 gives a calculated rise of 447 m.

The typical value of v in modern tall stacks is in the range of 60-120 fps, which is up from former stack designs by a factor of 3 or so. The high velocity allows a smaller stack to obtain a given amount of jetting. The desirable upper limit on v for head loss purposes is about 140 fps, but a much lower velocity induces mixing above the stack and limits the usual design to ≤ 105 fps.

The lower limit for v is the critical velocity. As the wind speed approaches the exit velocity, a point is reached where the plume is sheared off at the top of the stack and no plume rise occurs; this point is termed the critical velocity. The use of high exit velocities has led to positive stack pressures and more fan requirements. Positive pressures have resulted in sulfation problems in columns, necessitating repair or change of brick liners.

Stack exit temperatures (T_s) are normally in the range of 225°-350°F. The temperature is limited on the low side by the acid dewpoint of the stack gas and on the high side by the value of heat and the economics of its recovery.

The theoretical draft of the stack is the difference between the weight of the flue gas in the stack and that of an equal height of ambient air. The theoretical draft may be calculated from

$$SD_t = 0.256\,H_b p(1/T_a - 1/T_s) \,, \qquad (15\text{-}6)$$

where SD_t = the theoretical draft (in. w.g.), H_b = height of stack above breeching (ft), p = barometric pressure (in. Hg), and T_a and T_s = temperatures of ambient air and stack gas ($^{\circ}$R).

The available draft is the theoretical draft minus the friction losses in the stack. The friction loss is calculable from

$$h_L = 0.0942\,T_s\!\left(\frac{Q_m}{10^5}\right)^2 [\,1/D^4 + fH_b/D_a^{\,5}\,] \,, \qquad (15\text{-}7)$$

where h_L = frictional head loss (in. w.g.), Q_m = stack emission rate (lb/hr), D = exit diameter of stack (ft), D_a = average diameter of stack (ft), and f = friction factor [see Fig. 15-4, N_{Re} = $27,600\,Q_m/(T_s,D)$].

C. Structural Design

Stacks have been made of various materials--mostly steel, brick, and concrete. Tall stacks usually have an outer structural

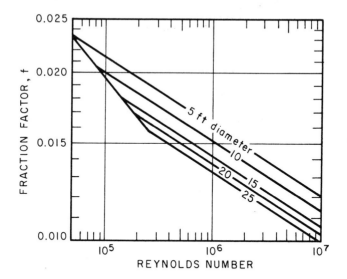

FIG. 15-4. Friction factors for stacks. From J. H. Perry
et al., Eds., Perry's Chemical Engineers' Handbook, 4th ed.,
McGraw-Hill, New York, 1963.

shell (the column) of reinforced concrete and an inner liner of
brick, steel, concrete, or other material. The very tall stacks
being built today have liners of corrosion-resistant steel, often
stainless steel near the top of the stack. The liners are usually
free-standing, but sometimes they are supported by the column,
especially the very tall stacks (see Fig. 15-5). The column
gives the liner protection from the wind and from some corrosion
and provides access to the liner, even while the stack is in use
with some designs. The annular clearance between the stack and
the liner is normally a minimum of 2.5 ft. The liner is likely to
be brick for stack heights less than 500 ft and steel for those
greater than 600 ft.

The width of the duct breeching(s) to the stack should be
no more than one-half of the liner inner diameter and no more than
one-third of the column outer diameter. If two openings are used,
they should be not more than 120° apart.

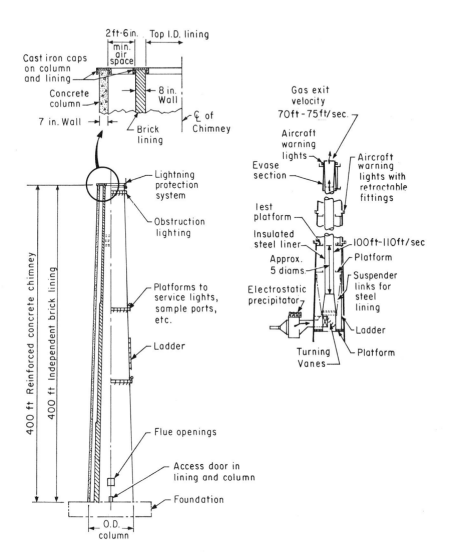

FIG. 15-5. Tall stack designs. (a) Free-standing liner
and (b) suspended liner. (a) From D. Carlton-Jones and H. B.
Schneider, "Tall Chimneys," Chem. Engr., 75:22, 166-169,
October 14, 1968. (b) From J. H. Perry et al., Eds., Perry's
Chemical Engineers' Handbook, 4th ed., McGraw-Hill, New
York, 1963.

A tall stack requires a strong foundation. A rule of thumb for a soil-bearing capacity of 2 tons/ft^2 is an octagonal base with a diameter equal to 1/10th the stack height plus the inside diameter of the liner at its top. The foundation size is increased about 15% in areas of earthquakes. The mortar selected depends on the stack conditions, but it is usually potassium silicate which is acid resistant and can withstand high temperatures. Slip-form construction of the column is difficult because of the taper of the stack; however, it is often used to save time. Sometimes only the bottom section of the stack is lined in order to save money (10-20%), but this is not usually good practice.

Parabolic nozzles have been put on the top of existing stacks to add effective stack height through higher jetting velocities (also increases head) and they have been added to new stacks with plans for removal on future plant expansion. Multiple flues are being put into a single column to provide for separate combustion units with near constant exit velocities. These liners extend up to several feet above the column to prevent downwash of the plume behind the stack.

IV. APPLICATIONS

The tall stack can be used to dilute any gas or aerosol stream. Low stacks (on the order of 200 ft high) have been used for dispersing many toxic, malodorous, and nuisance materials. Tall stacks (greater than about 400 ft) have been applied almost exclusively to the dilution of sulfur dioxide. The dilution of the particles and other gases accompanying the sulfur dioxide is an added benefit.

The ideal location for a tall stack is a hilltop; however, many combustion processes which use tall stacks also use large volumes of cooling water, and water pumpage costs dictate a valley location for the plant. There have been a few cases in

which plants were able to locate at the foot of a bluff, put the
stack on the bluff, and extend the ducts up to the stack.

Stack heights are continually increasing. Smelters of
pyritic ores have relatively high concentrations (several thousand
ppm_v) of sulfur dioxide in their stack gases. Inco (Sudbury,
Ontario) has a newly built, 1250 ft-high stack. Power plants
have larger volumes than smelters but with lower concentrations
of sulfur dioxide, usually less than a few thousand ppm_v. Power
plants are more likely to be located in deep valleys and require
stacks of about the same heights as do the smelters. American
Electric (Moundville, West Virginia) has a 1206-ft stack and an
826-ft stack (Brilliant, Ohio); Penelec (Conemaugh, Pennsylvania)
has two 1000-ft stacks; Georgia Power (Cartersville, Georgia)
has a 1000-ft stack with two flues (86-ft column diameter at the
bottom and 61 ft at the top); and TVA (Bull Run, Tennessee) has
an 800-ft stack (column 65 ft o.d. at the bottom and liner 28 ft
i.d. at the top). ASARCO (El Paso, Texas) has an 846-ft stack
on its copper smelter. A recent survey of power plant stacks
shows a current relation of stack height (H) to size of plant
(P, megawatts) of

$$H = 181 + 0.6P \text{ ft.} \qquad\qquad (15\text{-}8)$$

V. COSTS

The cost of a tall stack may well be a significant portion
of the plant cost. The tallest stack which has been built to date,
1250 ft high, cost some \$6.5 million. A rough estimate for the
cost of a 24 ft-diameter tall stack may be made using the relation

$$C = 30\,H^2 \quad \text{(for } H \geq 4\text{)}, \qquad\qquad (15\text{-}9)$$

where C = cost (1000's of dollars) and H = stack height (100's of
ft).

The operating costs for a tall stack will remain high even
if the maintenance costs are low. The pumpage costs are quite

large. For example, a new lignite-burning power plant at Fair-
field, Texas (Texas Utilities) has two 3500 HP fans for H = 400
ft, v = 90 fps, d = 22.2 ft, and T_s = 335°F on a 575-MW plant.
Fans are also used to furnish combustion air to the boiler.

VI. SUMMARY

Tall stacks dilute stack gases to acceptable ground level
concentrations. They are used primarily for sulfur dioxide from
solid fossil-fueled power generating stations and smelters of
pyritic ores.

The usual tall stack is a reinforced concrete outer shell
with a free-standing liner which will probably be of masonry if
the height is less than about 500 ft and of steel if the height is
greater than 600 ft. There may be more than one flue in a single
column for the purpose of obtaining more uniform stack velocities.

The first cost of tall stacks is considerable (about $1200
per foot for a 400 X 24 ft stack. The high first cost and the continued high operating costs for a tall stack justify a major effort
in the stack design and selection of the design parameters. The
rules of thumb that worked reasonably well for smaller stacks are
no longer adequate. These include such things as: The minimum
H should be 2.5 times the height of any surrounding building or
terrain; a 1°F rise in exit temperature increases ΔH by 2.5 ft;
and the days for erection may be calculated by H/4.5.

The development of better sulfur removal from solid fuel by
gasification and scrubbing and from pyritic ores by hydrometallurgy
coupled with increasingly stringent regulations and the rising
costs for tall stacks may soon end the construction of tall stacks.

PROBLEMS

1. Plot the downwind ground concentration (X) versus distance
    (x) for type D stability with a wind velocity (u) of 3.8 m/sec,
    an emission rate (Q) of 20,000 lb/hr of sulfur dioxide, and
    an effective stack height (H_e) of 400 m.

2. Fit a straight line to type D stability plots in Fig. 15-2; i.e., find a and b in $\sigma_y = ax^b$ and f and g in $\sigma_z = fx^g$. Use segments if necessary.

3. Obtain expression for distance to maximum ground concentration by substituting ax^b and fx^g from Problem 2 into Eq. 15-1, taking derivative (dX/dx), setting equal to zero, and solving for x.

4. What is the calculated cost per year for the heat added to get the 26-m plume rise in Example 15-2 if fuel is $\$0.75/10^6$ Btu?

5. What is the calculated annual cost for changing the H_e by 26 m through changing v, including additional frictional loss, when

$$\Delta H = [1.5(7.32)v + 1230]/3.8$$

6. Check the applicability of the rule of thumb that $1^\circ F$ gives 2.5 ft of plume rise with Example 15-2. What causes the discrepancy?

7. What is the effective stack height by the TVA method of the lignite power plant stack described above for neutral conditions and a wind speed of 3.8 m/sec?

8. What is the expected maximum ground level concentration for the stack in Problem 7 for type D stability if the plant burns 811,000 lb/hr of lignite with 0.8% sulfur? Where does the maximum concentration occur?

BIBLIOGRAPHY

Briggs, G. A., Plume Rise, TID-25075, U. S. Atomic Energy Commission, Oak Ridge, Tennessee, 1969.

Carlton-Jones, D., and H. B. Schneider, "Tall Chimneys," Chem. Engr., 75:22, 166-169, October 14, 1968.

Gifford, F. A., and D. H. Pack, "Surface Deposition of Airborne Material," Nuclear Safety, 3:4, 76-80, 1962.

Montgomery, T. L., S. B. Carpenter, W. C. Colbaugh, and F. W. Thomas, "Results of Recent TVA Investigations of Plume Rise," J. Air Pollution Control Assoc., 22:10, 779-784, October 1972.

Nelson, F., and L. Shenfeld, "Economics, Engineering and Air Pollution in the Design of Large Chimneys," J. Air Pollution Control Assoc., 15:8, 355-361, August 1965.

O'Connor, J. R., and J. F. Citarella, "An Air Pollution Control
    Cost Study of the Steam-Electric Power Generating Industry,"
    J. Air Pollution Control Assoc., 20:5, 285-288, May 1970.

Pasquill, F., Atmospheric Diffusion, Van Nostrand, New York,
    1962.

Perry, J. H., R. H. Perry, C. H. Chilton, and S. D. Kirkpatrick,
    Eds., Perry's Chemical Engineers' Handbook, 4th ed.,
    McGraw-Hill, New York, 1963.

Sporn, P., et al., The Tall Stack--For Air Pollution Control on
    Large Fossil-Fueled Power Plants, American Electric Power
    Co., New York, 1967.

Stephens, N. T., and R. O. McCaldin, "Attenuation of Power
    Station Plumes as Determined by Instrumented Aircraft,"
    Environmental Sci. Tech., 5:7, 615-621, July 1971.

Thomas, F. W., S. B. Carpenter, and F. E. Gartrell, "Stacks--
    How High?" J. Air Pollution Control Assoc., 13:5, 189-
    205, May 1963.

Turner, D. B., Workbook of Atmospheric Dispersion Estimates,
    NCAPC, U. S. Public Health Service Publ. No. 999-AP-26,
    Cincinnati, Ohio, 1967.

Section III: Gaseous Control

Air pollution control often involves the removal of a polluting gaseous component from a gas stream, or the conversion of the component into innocuous or less offensive substances, before the gas stream is vented to the atmosphere. The gaseous component may be removed by absorption into a liquid, adsorption onto a solid, or, more rarely, condensation into a liquid. The liquid or solid may be disposed of or the pollutant may be removed in a concentrated form for disposal and the absorbent or adsorbent regenerated for re-use. It is especially important that no other pollution problem be created during the disposal. Gaseous components that produce relatively inoffensive oxidation products are often incinerated.

Chapter 16

ABSORPTION

I. INTRODUCTION

Gases are frequently dissolved (dispersed on a molecular basis) in liquids, absorbed, for the purpose of air pollution control. The gas stream is then vented to the atmosphere, and the liquid containing the polluting gas may be disposed of or the sorption process may be reversed to desorb the gas in a concentrated form for disposal or use. The gas may be absorbed in a reactive liquid to form a useful compound or one which is more readily separable for disposal; sometimes the compound is precipitated from the liquid. The processes of liquid-liquid absorption (solvent extraction) and solid-liquid dissolution and liquid-gas desorption transport are not described in detail here, but the methods for these processes may be inferred from the coverage of gas-liquid absorption. The usual absorption process in air pollution control is to pass the polluted airstream up through a packed tower which contains a large surface matrix (packing) wetted with water; however, venturi scrubbers, spray towers, and bubble plate towers are used in nearly 10% of the applications (see Fig. 16-1).

II. THEORY

When a gas is placed in contact with a liquid, the diffusion process causes gas molecules to enter the liquid, i.e., absorption occurs. If the gas (absorbate) does not react chemically with the liquid (absorbent), there will be diffusion of the entrained gas out of the liquid; i.e., desorption takes place. Operation of a removal process obviously requires that the absorption rate be

99

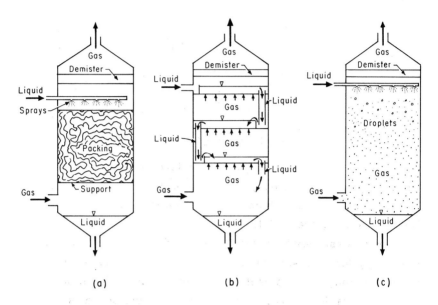

FIG. 16-1. Types of absorption towers. (a) Packed tower, (b) tray or plate tower, and (c) spray tower.

greater than the desorption rate. When the rates are equal, equilibrium exists. A plot of the gaseous pressures versus their equilibrium concentrations is called an equilibrium curve (see Fig. 16-2). (Designated c and p values are defined later.)

Slightly soluble gases follow Henry's law which states that the equilibrium plot is a straight line or

$$p = Hx, \qquad (16-1)$$

where p = pressure of the absorbing component (mm Hg), x = mole fraction of absorbing component in liquid phase, and H = Henry's constant (mm Hg). Even the very soluble gases may be assumed to follow Henry's law at the dilute concentrations usually found in air pollution control applications. Table 16-1 shows values of Henry's constants for some common gases.

TABLE 16-1

Henry's Law Constants for Gases in Water. (10^{-4} H atm/mole fraction)

Temperature (°C)	Air	CO_2	CO	H_2S	NO	N_2	O_2	SO_2	NH_3	COS
0	4.32	0.0728	3.52	0.0268	1.69	5.29	2.55	0.0011	0.000034	0.092
10	5.49	0.104	4.42	0.0367	2.18	6.68	3.27	0.0017	0.000054	0.148
20	6.64	0.142	5.36	0.0483	2.64	8.04	4.01	0.0024	0.000090	0.219
30	7.71	0.186	6.20	0.0609	3.10	9.24	4.75	0.0034	0.00013	0.304
40	8.70	0.233	6.96	0.0745	3.52	10.4	5.35	0.0054	0.00019	--
50	9.46	0.283	7.61	0.0884	3.90	11.3	5.88	0.0070	0.00029	--
60	10.1	0.341	8.21	0.103	4.18	12.0	6.29	--	0.00038	--
70	10.5	--	8.45	0.119	4.38	12.5	6.63	0.013	--	--
80	10.7	--	8.45	0.135	4.48	12.6	6.87	--	--	--
90	10.8	--	8.46	0.144	4.52	12.6	6.99	--	--	--
100	10.9	--	8.46	0.148	4.54	12.6	7.01	0.026	--	--

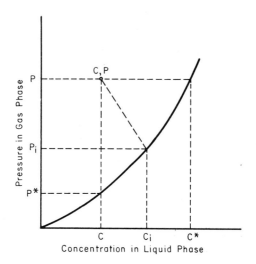

FIG. 16-2. Equilibrium curve.

Material in the body of a turbulent fluid is transported by
two processes, eddy diffusion and molecular diffusion. The most
popular approach to explaining the transport of materials across
a phase interface between a gas and a liquid is the two-film theory.
This theory holds that stagnant boundary layers of both the gas
and the liquid exist at the interface (see Fig. 16-3), that diffusion across these boundary layers is purely molecular in nature,
and that the transport rate across the boundary layers is so slow
that transport in the bodies of the fluids can be ignored.

Molecular diffusion occurs in the direction of the negative
concentration gradient, and driving forces for the transport are
as shown in Fig. 16-2. The rate of absorptive transport across
the gas boundary must equal that across the liquid,

$$r = K_G aV(p - p_i) = K_L aV(c_i - c),$$ (16-2)

where r = absorption rate (lb mole/hr), K_G = overall gaseous
transport constant (lb mole/hr-ft^2-atm), a = absorbing area per

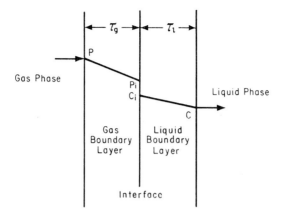

FIG. 16-3. Two-film concept of absorption.

unit volume of absorber (ft^2/ft^3), V = volume of absorber (ft^3), p and p_i = gas pressures for absorbate in body of gas and at interface, respectively (atm), K_L = overall constant for liquid transport (ft/hr), and c_i and c = concentrations of absorbate at interface and in body of liquid, respectively (lb mole/ft^3).

Figure 16-4 shows diagrammatic sketches of absorber tower-operating lines for countercurrent and cocurrent modes of operation. The parameters at the bottom of the tower are conventionally identified with the subscript 1 and those at the top of the tower with the subscript 2. The general terms at some point in the tower are designated without subscripts.

A materials balance across the entire countercurrent tower (between points 1 and 2) shows

$$G_1y_1 + L_2x_2 = G_2y_2 + L_1x_1 , \qquad (16-3)$$

where G = gas flow rate (lb mole/ft^2-hr), L = liquid flow rate (lb mole/ft^2-hr), y = mole fraction of absorbate in gas phase, and x = mole fraction of absorbate in liquid phase. If it is assumed that the system is dilute enough that G and L do not vary

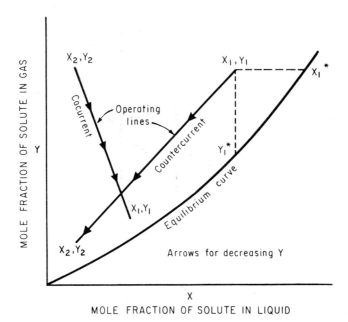

FIG. 16-4. Cocurrent and countercurrent absorption modes.

significantly over the height of the tower, then

$$G(y_1 - y_2) = L(x_1 - x_2) \text{ for countercurrent towers,} \quad (16\text{-}4)$$

$$G(y_2 - y_1) = L(x_1 - x_2) \text{ for cocurrent towers.}$$

Equations 16-4 are the equations for the operating lines (see Fig. 16-4). In general form

$$y = (L/G)x + y_2 . \quad (16\text{-}5)$$

Although the interface pressures and concentrations were used in defining the driving pressures in Eq. 16-2 and they should fall on the equilibrium curve in each case, it is impossible to know where on the curve the values are located; therefore, the values which would be in equilibrium with the amount in the opposite phase are used. These values are designated with asterisk superscripts, as shown in Figs. 16-2 and 16-4. The driving pressure and concentration become (p - p*) and (c* - c), and

Eq. 16-2 for the mole fraction basis can be written

$$r = K_G a \, VP(y - y^*) = K_L a \, V\rho_M (x^* - x), \qquad (16-6)$$

where r = rate of absorption (lb mole/hr), P = total pressure of
gas phase (atm), ρ_M = molar density of liquid phase (lb mole/ft^3),
and other terms are as previously defined.

The resistances to absorption may be written as

$$1/(K_G a) = 1/(k_G a) + m/(k_L a) \,, \qquad (16-7)$$

$$1/(K_L a) = 1/(k_L a) - 1/(k_G a m),$$

where K_G, K_L, and a are the same as before, k_G = transfer coefficient through the gas boundary film, k_L = transfer coefficient
through the liquid boundary film, and m = slope of the equilibrium
curve (line). For very soluble gases most of the resistance is in
the gas film, and the gas film controls the rate of absorption.
For very insoluble gases most of the resistance is in the liquid
film, and the liquid film controls the rate of absorption. In these
cases the controlling film resistance is computed as

$$1/(K_G a) \simeq 1/(k_G a) \text{ and } 1/(K_L a) \simeq 1/(k_L a). \qquad (16-8)$$

The heights of tower to overcome the resistances are
formulated for straight lines, operating and equilibrium, as

$$H_{OG} = H_G + H_L mG/L = H_{OL} mG/L \,, \qquad (16-9)$$

where H_{OG} = height of overall gas transfer unit (ft), H_G = height
of gas transfer unit (ft), H_L = height of liquid transfer unit (ft),
mG/L = absorption factor (see below), and H_{OL} = height of overall
liquid transfer unit (ft). The term mG/L is the ratio of the slopes
of the equilibrium line (m) and the operating line (L/G); it is
termed the absorption factor and recurs repeatedly in absorber
design.

III. DESIGN

Design of absorbers for air pollution control is largely
concerned with determination of the area of interfacial contact

needed and how to provide that area. The principal design decisions are the type of absorber tower (packed, spray, bubble-
plate, or other), the liquid flow rate (slope of operating line),
kind of solvent (generally water), operating pressure (usually
1 atm), type of packing, tower diameter, and tower height.

Although the tower may operate cocurrently or even cross-
currently, the emphasis here is placed on countercurrent operation
because of its overwhelming popularity. The equilibrium line is
assumed straight and the log mean pressure and concentration are
defined for the purpose of simplifying calculations.

A. Tower Diameter

The tower diameter is selected to give a gas velocity that
is satisfactory. The gas velocity in countercurrent towers must
be low enough to prevent excessive holdup and/or carryover of
the liquid, yet high enough for economical design.

In spray towers, the droplet size and the gas velocity are
chosen to give settling out of the drops as desired for achieving
absorption by the area of the drops and the contact time. If
particulate removal is to be accomplished in the same spray tower,
the relative velocity of the gas and the droplet needs to be rather
high, especially for small particles. For absorptive gas removal
only, drops large enough to settle without carryover (allowing
for evaporative decrease in size) are required.

The design diameters of packed and plate or tray towers
may be set from a consideration of the flooding velocity. The
design velocity is taken as a percentage of the flooding velocity,
usually 60-70% for packed towers and up to 80% for plate or tray
towers. The theoretical minimum L/G, that ratio which would
result in a saturated liquid from the amount of gas absorbed, is
increased by 25-100% for a practical design. Figure 16-5a can

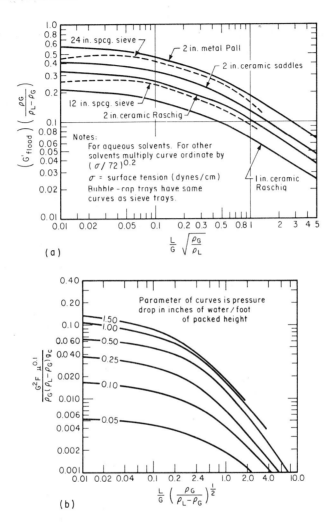

FIG. 16-5. Gas velocity determinations. (a) Flood velocities and (b) pressure drops. (L = liquid rate, lb/sec-ft^2; G = gas
rate, lb/sec-ft^2; ρ_L = liquid density, lb/ft^3; ρ_G = gas density,
lb/ft^3; F = packing factor; μ = viscosity of liquid, centipoises;
and g_c = gravitational constant = 32.2 ft/sec^2.) From J. R. Fair,
"Sorption Processes for Gas Separation," Chem. Engr., 76:15,
90-110, July 14, 1969 and from Design Information for Packed
Towers, Bulletin DC-10R, Norton: Chemical Process Products
Division, Akron, Ohio, 1971.

then be used to obtain the superficial gas velocity to cause
flooding. The tower diameter required is

$$D = 4 Q/(\pi U_G)^{0.5} , \qquad\qquad (16\text{-}10)$$

where D = tower diameter (ft), Q = gas flow rate (cfs), and U_G
= superficial gas velocity (ft/sec) = Q/tower area in ft^2. J. R.
Fair warns that foaming may greatly reduce the U_G allowed and
antifoaming agents may be needed.

The tower diameters for packed towers are often determined
by limiting the pressure drop to some maximum desirable design
value and using Fig. 16-5b. This method was developed by
Sherwood and modified by Eckert who substituted the packing
factor for a/ε^3 (packing area per unit volume/voids ratio cubed).
Table 16-2 shows packing factors for some of the popular tower
packings that are depicted in Fig. 16-6. Pressure drops for

BERL SADDLE

RASCHIG RING

INTALOX SADDLE

PALL RING

TELLERETTE

FIG. 16-6. Some popular packings. From Control Tech-
niques for Hydrocarbon and Organic Solvent Emissions from
Stationary Sources, National Air Pollution Control Administration
Publ. No. AP-68, Washington, D. C., March 1970.

TABLE 16-2

Design Constants for Ceramic Tower Packings

Packing Type	Size (in.)	Packing factors $F \simeq ac^3$	ε (%)	Empirical design constants α	β	γ	Φ	η	m	n	$K_G a$ (CO_2:4% NaOH) (lb mole/hr-ft^3-atm)
Raschig rings	1	155	73	6.41	0.32	0.51	0.100	0.22	32.10	0.00434	2.2
	2	65	74	3.82	0.41	0.45	0.0125	0.22	11.13	0.00295	1.6
Berl saddles	1	110	--	1.97	0.36	0.40	0.00588	0.28	16.01	0.00295	--
	2	45	--	5.05	0.32	0.45	0.00625	0.28	8.01	0.00225	--
Intalox saddles	1	98	78	--	--	--	--	--	12.44	0.00277	2.6
	2	40	79	--	--	--	--	--	--	--	2.0

packed absorption towers are usually in the range of 0.2-0.4 in. w.g./ft of packing for water as the absorbent and are lower or higher for lighter or heavier absorbents. The abscissa value for use with Fig. 16-5b is obtained as above, the head loss is selected, and the ordinate value obtained. All the parameters in the ordinate value except the G are known; therefore, G can be calculated and the diameter of the tower determined.

EXAMPLE 16-1: Determine the diameter of a packed tower with 2-in. ceramic Raschig rings to take out (using water) 90% of the 5%$_v$ ammonia present in 4000 cfm of gas (air and ammonia) at 55°C. The maximum tower temperature will be about 40°C.

Minimum liquid flow: Minimum flow = equilibrium flow

Solubility @ 40°C: 0.05 = 1.92x   x = 0.026 (Table 16-1)

$(X/17)/(100/18 + X/17) = 0.026$ $X = 2.5$ lb NH_3/100 lb H_2O

NH_3 removed $= 0.9(0.05)(4000)[17(293)/(385 \times 328)]$
$\qquad = 7.10$ lb/min

Minimum water $= 7.10/(2.52/100) = 282$ lb/min

Design liquid flow: Use 425 lb/min ($\sim$1.5 theoretical)

$$\frac{[\rho_G/(\rho_L - \rho_G)]^{0.5}}{}:$$

Average $NH_3 = (0.05 + 0.005)/2 = 0.028$

$\rho_G = [0.028(17) + 0.972(29)]/(385 \times 313/293) = 0.0697$
$\qquad\qquad\qquad\qquad\qquad\qquad\qquad\qquad\qquad\qquad$ lb/ft^3

$[0.0697/(62.4 - 0.0697)]^{0.5} = 0.0334 \approx \rho_G/\rho_L$

$L/G = (425)/[0.0697(313/328)(4000)] = 1.60$

$L/G\sqrt{(\rho_G/\rho_L)} = 1.60(0.0334) = 0.0534$

Diameter by flooding velocity:

Fig. 5a: $G'_f\sqrt{[\rho_G/(\rho_L - \rho_G)]} = 0.285$

$G'_f = 0.285/0.0334 = 8.53$ ft/sec $G' = 60\% G'_f = 5.12$ ft/sec

$Q = [4000 - 0.9(0.05)(4000)/2]313/328 = 3730$ cfm @ 40°C

$D = [4Q/(G'\pi)]^{0.5} = [4(3730/60)/(5.12\pi)]^{0.5} = 3.93$ ft

Diameter by head loss:

Fig. 5b: Use head loss = 0.35 in. w.g./ft of packing

$G' = 3730(0.0697) = 260$ lb/min

$$G^2 F\mu^{0.1} / [g\rho_G(\rho_L - \rho_G)] = 0.040 \text{ for abscissa of } 0.0534$$

$$G = [0.040(0.0697)(62.4 - 0.0697)(32.2)/(65 \times 0.656^{0.1})]^{0.5}$$

$$= 0.300 \text{ lb/sec-ft}^2$$

Area required $= (260/60)/0.300 = 14.43 \text{ ft}^2$

Diameter $= (4A/\pi)^{0.5} = (4 \times 14.43/3.14)^{0.5} = 4.29 \text{ ft}$

B. Tower Height

Once the diameter of the tower has been set, the height of
the tower (packing) must be determined to obtain the desired
efficiency of absorption. The basic design parameters are the
area of contact and the time of contact, both of which, for a
given medium, are related to the volume of the tower.

For spray towers the height required is that height necessary
to provide the droplet area and the residence time. The area of
the droplets is calculated as

$$a = na' = [Lt/(\pi D^3/6)]\pi D^2 = 6 Lt/D , \qquad (16\text{-}11)$$

where a = total absorption area of droplets, n = number of droplets
in the tower at one time, a' = area of a single droplet, D = diameter of droplet, and t = holdup time for droplets = $Z/(u_t - v_g)$,
Z = height of tower, u_t = terminal settling velocity of droplets,
and v_g = velocity of gas relative to tower (up positive).

For packed towers, the absorption across a differential
height of tower is formulated in terms of the gas and liquid phases,
respectively, as

$$Z = \frac{G}{K_G a} \int_{p_1}^{p_2} \frac{dp}{(P - p)(p - p^*)} , \qquad (16\text{-}12)$$

$$Z = \frac{L}{K_L a} \int_{c_1}^{c_2} \frac{dc}{(\rho_M - c)(c^* - c)} ,$$

where Z = tower packing height and all other terms are as previously defined. These equations are solvable by graphing the
equilibrium curve and operating line, determining the appropriate

values from the graph, and numerically integrating the functions under the integral signs. For example, $[(P - p)(p - p^*)]^{-1}$ is plotted versus p for eight or ten step values. The $K_G a$ and $K_L a$ values are obtained from pilot plant data on the same type of installation.

For the dilute systems usually encountered in air pollution control, $p \ll P$ and $c \ll \rho_M$; therefore, Eqs. 16-12 may be simplified to

$$Z = \frac{G}{K_G aP} \int_{p_1}^{p_2} \frac{dp}{p - p^*} , \qquad (16\text{-}13)$$

$$Z = \frac{L}{K_L a \rho_M} \int_{c_1}^{c_2} \frac{dc}{c^* - c} .$$

The height of the packed tower is calculated from

$$Z = H_{OG} N_{OG} \quad \text{or} \quad Z = H_{OL} N_{OL} , \qquad (16\text{-}14)$$

where H_{OG} = height of an overall gas transfer unit = $G/(K_G aP)$, N_{OG} = number of gas transfer units = integral term in pressure equation, H_{OL} = height of an overall liquid transfer unit = $L/(K_L a \rho_M)$, and N_{OL} = number of liquid phase transfer units = integral term in concentration equation.

1. H_{OG} or H_{OL} Determination

The best determinations of H_{OG} or H_{OL} are made from pilot plant data on the same gas stream with $K_G a$ and $K_L a$ values. The $K_G a$ or $K_L a$ is used with the relation as defined in Eq. 16-14 for the calculation of H_{OG} or H_{OL}. If pilot plant data are unavailable, H_{OG} may be estimated by Eq. 16-9 through the use of the following relations:

$$H_G = \alpha (G^\beta / L^\gamma)[\mu_G/(\rho_G \mathcal{D})]^{0.5} , \qquad (16\text{-}15)$$

$$H_L = \phi (L/\mu_L)^\eta [\mu_L/(\rho_L \mathcal{D})]^{0.5} ,$$

where H_G and H_L are in ft, G and L are in lb/hr-ft^2, μ_L is in

lb/ft-hr in the first term, and the viscosity, density, and diffusivities in the gas and liquid phases are in units to give a
dimensionless number (Schmidt number) in the bracketed terms.
(See Table 16-2 for some values of parameters α, β, γ, Φ, and η.)

<u>EXAMPLE 16-2</u>: Find the H_{OG} for the problem described in
Example 16-1 if D = 4.25 ft. (A = 14.2 ft^2)

$H_{OG} = H_G + (mG_M/L_M)\,H_L$ (without pilot plant K_Ga, use estimating formula)

m = y*/x = 0.0592/0.0308 = 1.92 (Table 16-1)

G = 260(60)/14.2 = 1100 lb/hr-ft^2

G_M = 1100/[0.028(17) + 0.972(29)] = 38.4 lb mole/hr-ft^2

Total entering gas = 4000(60)/[385(328/293)(14.2)]

= 39.2 lb mole/hr-ft^2

L = [425(60) + 0.028(39.2)(17)]/14.2 = 1797 lb/hr-ft^2

L_M = [425(60)/18 + 0.028(39.2)]/14.2 = 99.8 lb mole/hr-ft^2

H_G = 3.82[1100$^{0.41}$/1797$^{0.45}$][0.00019/(0.00112 X 0.254)]$^{0.5}$

= 1.89 ft (for Diffusivities and Schmidt numbers, see Table 16-3)

H_L = 0.0125[1797/158.6]$^{0.22}$[0.00656/(1.00 X 2.88X10^{-5})]$^{0.5}$

= 0.322 ft

H_{OG} = 1.89 + 1.92(38.4/99.8)(0.322) = 2.13 ft

TABLE 16-3

Diffusivity Coefficients in Air and in Water

Solute	D_G in air @ 25°C/1 atm (ft^2/hr)	Schmidt No.	D_L in water @ 20°C (ft^2/hr)	Schmidt No.
Ammonia	0.90	0.67	7.3 X 10^{-5}	570
Carbon dioxide	0.64	0.94	5.8 X 10^{-5}	570
Oxygen	0.80	0.75	7.0 X 10^{-5}	558
Methanol	0.62	0.97	5.0 X 10^{-5}	785
Acetic acid	0.51	1.16	3.4 X 10^{-5}	1140

2. N_{OG} or N_{OL} Determination

There are at least several methods for determining the N_{OG} or N_{OL} value. Four of these methods will be presented here. The discussions are sometimes limited to the N_{OG} value in the interests of brevity, but the N_{OL} may be determined in a similar fashion.

a. Log Mean Method. One method of evaluating N_{OG} makes use of the log mean concept employed in heat transfer problems. When both equilibrium and operating plots are straight lines,

$$\int_{p_1}^{p_2} \frac{dp}{p - p^*} = \frac{p_1 - p_2}{(p - p^*)_{LM}} , \qquad (16\text{-}16)$$

where $(p - p^*)_{LM} = \dfrac{(p_1 - p_1^*) - (p_2 - p_2^*)}{\ln[(p_1 - p_1^*)/(p_2 - p_2^*)]}$ and

$$\int_{c_1}^{c_2} \frac{dc}{c^* - c} = \frac{c_1 - c_2}{(c^* - c)_{LM}} ,$$

where $(c^* - c)_{LM} = \dfrac{(c_1^* - c_1) - (c_2^* - c_2)}{\ln[(c_1^* - c_1)/(c_2^* - c_2)]}$.

EXAMPLE 16-3: Determine N_{OG} and height of tower packing (Z) for the problem used in Examples 16-1 and 16-2.

$$N_{OG} = \int_{p_1}^{p_2} \frac{dp}{p - p^*} = \frac{p_1 - p_2}{(p - p^*)_{LM}}$$

$$= \frac{(0.05 - 0.005)\ln[(0.05 - 0.0334)/(0.005 - 0.00)]}{(0.05 - 0.0334) - (0.005 - 0.00)}$$

$$= 4.66 \qquad [y^* = 0.0174(1.92) = 0.0334]$$

$$Z = 2.13(4.66) = 9.92 \text{ ft}$$

b. Baker Graphical Method. The N_{OG} can be obtained by the Baker graphical method. The first step is to draw a curve (A-A in Fig. 16-7) that goes through the midpoints of all possible vertical (p - p*) lines between the operating and equilibrium

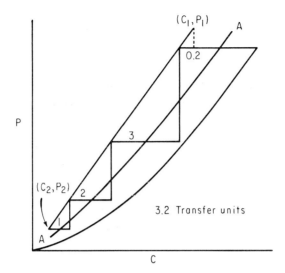

FIG. 16-7. Baker graphical technique.

curves. At point p_2,c_2 (the lower end of the operating line),
draw a horizontal line segment which is bisected by A-A, draw
a vertical from the right end of the line segment to the operating
line, then repeat the last two steps until the horizontal line segment crosses the c_1 value. Count the number of horizontal lines
and add the fraction of the final line needed to reach c_1. This
number is N_{OG}.

The Baker method can also be used to get N_{OL}. The procedure is to draw A-A such that it bisects the horizontal lines
between the equilibrium and operating curves $(c^* - c)$, then
starting at p_1,c_1 with vertical segments which are bisected by
A-A.

Considerable care must be exercised in using the Baker
method, especially when the curves are close together. Careful
constructions on a large scale graph with a sharp pencil are
warranted.

c. Theoretical Plate Method. The theoretical plate is a
hypothetical plate or tray which would bring the gas into equilibrium with the liquid on that plate in the absorption tower. The
number of theoretical plates can be determined graphically by a
stepping technique between the operating and equilibrium curves.
Starting at p_2, c_2 draw a horizontal line to the equilibrium curve,
draw a vertical line to the operating line, repeat the sequence
until c_1 is reached, and count the number of steps and add the
fraction of a step. The number obtained is the number of theoretical plates.

Because prohibitive time would be required to reach the
equilibrium on a plate, an actual plate is never equal to a theoretical plate in efficiency. The tower height is formulated as

$$Z = N_{TP} Z_{sp} / E ,$$ (16-17)

where Z = tower height (ft), N_{TP} = number of theoretical plates,
Z_{sp} = spacing of plates (ft), and E = overall absorption efficiency
on a plate.

d. Method for m near Zero. For very soluble gases and
for gases that react rapidly with the absorbent, the driving
pressure is approximately the total pressure of the solute; i.e.,
$p - p^* \simeq p$. In such cases

$$N_{OG} \simeq \ln(p_1 / p_2) .$$ (16-18)

In other words, N_{OG} is approximately the number of transfer
units (see Chapter 12). This approximation is equivalent to
assuming that the slope of the equilibrium plot is zero. For
hydrogen chloride, m is 0.0002 at $30^{\circ}C$ for molar concentrations
in the gas and liquid phases.

C. Head Loss

The head loss for the packed tower is selected to provide
an economic balance between the pumping costs and the tower

size. The velocity must be small enough to prevent excessive
holdup of the scrubbing liquid. The packing manufacturer supplies
design data that include the head loss per foot of packing. The
head loss for spray towers may be taken as the entry and exit
losses with the loss in the tower itself neglected if the tower flow
is straight. For cyclonic air flow the head loss in the spray tower
may be calculated as for the cyclone (see Chapter 19). The head
loss in tray towers may be calculated from

$$\Delta h = 0.186 (\rho_G/\rho_L)(U_0/C_0)^2 + [0.4 \exp(-0.92 \, U_0\sqrt{\rho_G}) + 0.6]h_w,$$
$$(16-19)$$

where Δh = pressure drop (in. w.g.), ρ_G and ρ_L = densities of
gas and liquid (both in same units), U_0 = velocity of gas through
openings (ft/sec), C_0 = orifice coefficient = 0.7-0.8 for sieve
trays = 0.6-0.7 for bubble cap trays, and h_w = height of weir
crest plus height of weir above tray (in.).

IV. APPLICATIONS

The design techniques described above need experienced
judgment in their application. Items which should receive careful consideration are entrainment and carryover of mists, foaming
or frothing, heat of absorption, and corrosion or scaling of tower.
Mist elimination is often carried out by placing an impactive mist
collector after the absorption portion of the tower. Foaming can
usually be controlled with an antifoaming agent added to the
absorbent. Heat of absorption is seldom a problem with the
dilute concentrations found in air pollution control, but it may
be removed where necessary with internal cooling in the tower.
Corrosion problems are prevented by the judicious choice of
tower materials--ceramics, stainless steel, and plastics may be
used throughout or as coatings on other materials. Scaling may
be controlled by adding chemicals to the absorbent. If scaling
is difficult to control, plate towers or even spray towers should
be given preference over packed towers.

Most absorption for air pollution control has been done in countercurrent towers, the exceptions being scrubbers designed primarily for small particle removal but serving the dual purpose of absorption. The highly soluble gases may often be removed more economically by cocurrent or crossflow absorbers because of the lower head losses in these types of units.

For very soluble gases, the gas film controls the rate of absorption, and, theoretically, spray droplets should be used to shear the gas boundary layer and reduce its thickness (τ_G) as much as possible. Conversely, for very insoluble gases, the liquid film controls the rate of absorption and minimum liquid boundary layer thickness (τ_L) and maximum rate of absorption would be obtained by bubbling the gas through the liquid. The normal practice is to use a packed tower in either case unless other considerations control the selection. Spray towers are not usually applicable to obtain high efficiency of removal because of the very large tower size required for a transfer unit.

Current water pollution regulations practically preclude the once-through use of the absorbent. Recirculation of the absorbent with regeneration or neutralization for disposal must be practiced.

Although water is the most common solvent that is used for absorption, many other solvents have been used. These include oils, oxidizing agents, acid and alkali solutions, mono-, di-, and tri-ethanol amine (especially for hydrogen sulfide), aniline for sulfur dioxide, and solvents containing perfume or odor-masking agents. The solvent that is used should readily absorb the solute with little evaporation, and be inexpensive, noncorrosive, and relatively nonflammable.

Absorption for air pollution control has been used to absorb gases and vapors and some particulates in cleaning sour natural gas and the off-gases from rendering, coffee roasting, fish meal processing, paper making, meat smoking, tar and asphalt operations, and metal plating.

The trend in packed towers has been toward higher flow
rates for both the gas and liquid phases. Some new shapes of
packing have appeared in the last several years, especially the
plastic packings. Various sizes of packing are available; a
general rule seems to be that packings 1 in. and smaller are
limited to towers 1 ft or less in diameter and 2-3 in. packings
are limited to towers with diameters of 3 ft or more.

The gas and liquid flows must be distributed evenly over
the cross section of the tower for best operation. The packing
must be supported on a strong base. Packing height is normally
limited to about 20 ft or 8 diameters of column, whichever is less.

V. COSTS

Absorbers in the 5000-10,000 cfm size range may cost from
less than $1/cfm for a simple spray tower to more than $10/cfm
for a packed tower with a stainless-steel shell. The costs vary
widely with efficiency and difficulty of removal. A packed column
in a carbon steel shell will likely cost $4-5/cfm and the bubble
cap column about $3-4/cfm. Cost estimates are sometimes based
on incremental height of tower. The cost of a 7500-cfm tower
may be $100-400/ft of height for gas flow rates of 300-500 cfm/ft^2.

VI. SUMMARY

Soluble gaseous air pollutants may be removed by absorption in a liquid. The transport rate into the liquid is strongly
dependent on the area of contact of the gaseous and liquid phases;
therefore, the process design is based on giving a large area of
contact. This is accomplished through a spray of liquid droplets
in the gas, bubbles of gas in the liquid, or a film of liquid on a
large area-to-volume matrix.

The driving force in absorption is the difference between
saturation concentration in the liquid and the actual concentration
present. The absorption of the not too soluble gases may be en-

hanced by reacting the dissolved gas to lower its concentration in the absorbent.

The absorbent may be regenerated by changing the pressure to cause a release of the absorbed gas or by chemical treatment or it may be neutralized for disposal. Sometimes the solution is valuable as it is produced.

PROBLEMS

1. Work the problem in Example 16-1 using 2-in. berl saddles instead of Raschig rings.

2. Work the problem in Example 16-2 using 2-in. berl saddles instead of Raschig rings.

3. Work the problem in Example 16-3 using 2-in. berl saddles instead of Raschig rings.

4. What is the $K_G a$ for the conditions in Examples 16-1, 16-2, and 16-3?

5. Design a packed tower to reduce the sulfur dioxide in 2 million acfm of stack gas at $300°F$ from 2400 ppm_v to 400 ppm_v. Is this design practical? Why?

6. Describe the alkaline absorption of sulfur dioxide in a marble bed scrubber.

7. What are the advantages of recirculation of the scrubbing liquor in ammonia absorption?

8. Describe the absorption of hydrogen sulfide from sour natural gas with a mono-ethanol amine solution. Note particularly the pressures involved in absorption and desorption cycles.

BIBLIOGRAPHY

Air Pollution Manual, Part II: Control Equipment, American Industrial Hygiene Assoc., Detroit, Michigan, 1968.

Calvert, S., "Source Control by Liquid Scrubbing," Air Pollution, Vol. III (A. C. Stern, ed.), 2nd ed., Chapter 46. Academic Press, New York, 1968.

Control Techniques for Hydrocarbon and Organic Solvent Emissions from Stationary Sources, National Air Pollution Control Administration Publ. No. AP-68, Washington, D. C., March 1970.

Danielson, J. A., Ed., Air Pollution Engineering Manual, NCAPC,
   U. S. Public Health Service Publ. No. 999-AP-40, Cincinnati, Ohio, 1967.

Design Information for Packed Towers, Bulletin DC-10R, Norton:
   Chemical Process Products Division, Akron, Ohio, 1971.

Fair, J. R., "Sorption Processes for Gas Separation," Chem.
   Engr., 76:15, 90-110, July 14, 1969.

Leva, M., Tower Packings and Packed Tower Design, 2nd ed.,
   U. S. Stoneware Co., Akron, 1953. (Now Norton above)

Liddle, C. J., "How to Design Desorption Systems Based on
   Pressure Reduction," Chem. Engr., 77:15, 87-94, July 13,
   1970.

Perry, J. H., R. H. Perry, C. H. Chilton, and S. D. Kirkpatrick,
   Eds., Perry's Chemical Engineers' Handbook, 4th ed.,
   McGraw-Hill, New York, 1963.

Ranz, W. E., "Source Control by Liquid Scrubbing," Air Pollution,
   Vol. II (A. C. Stern, ed.), 1st ed., Chap. 31, Academic
   Press, New York, 1963.

Rich, L. G., Unit Operations of Sanitary Engineering, Wiley,
   New York, 1961.

Sherwood, T. K., Absorption and Extraction, 1st ed., McGraw-
   Hill, New York, 1937.

Strauss, W., Industrial Gas Cleaning, Pergamon Press, New
   York, 1966.

Teller, A. J., "Selection of Air Pollution Control Equipment,"
   Engr. Prog. at Univ. of Fla., XV:9, September 1961.
   Reprinted from Ind. Water Wastes, January-February 1961.

Treybal, R. E., Mass-Transfer Operations, McGraw-Hill, New
   York, 1955.

Zenz, F. A., "Designing Gas-Absorption Towers," Chem. Engr.,
   79:25, 120-138, November 13, 1972.

Chapter 17

ADSORPTION

I. INTRODUCTION

Adsorption is a surface phenomenon by which materials in gaseous mixture or liquid solution are concentrated at the interface with a nonmixing substance--for air cleaning purposes the collection moves from the gaseous to the solid surface. Adsorption is widely used to separate gaseous materials from airstreams. Despite the concentration dependency of adsorption, the process finds successful applications in the removal of very dilute components, especially odors. Perhaps its most frequent application in air pollution control at the present is in solvent recovery.

II. THEORY OF ADSORPTION

Adsorption is usually physical in nature with the gaseous material (adsorbate) essentially condensing on the solid (adsorbent). The reaction is exothermic, the molecules giving up their kinetic energy--practically the energy of liquefaction. Physically adsorbed material may be driven off the adsorbent in its preadsorbed form by simple heating for vaporization. Chemisorption occurs if the material reacts on the adsorbent and is not recoverable as the original material. Chemisorption and physical adsorption often occur concurrently and give an energy release in excess of that for physical adsorption alone, usually two to three times the energy for liquefaction.

Efficient adsorbent particles are characterized by very large surface areas (perhaps as much as 1000 m^2/gm, or even more). The large surface area is mostly a result of the porous structure of the particles. Adsorption in the depths of the pores is by

123

migration of the adsorbate which is originally adsorbed at the
exterior of the particle. Because of the migration, rapidly saturated adsorbent may develop additional capacity on storage.

There have been many hypotheses about the actual mechanism of adsorption and whether the adsorbate is in a monolayer
or a multilayer on the adsorbent. Fortunately, knowledge of the
forces of adsorption and the number of layers of adsorbate is not
essential to the empirical design of adsorptive air cleaners.

III. DESIGN OF ADSORBERS

Adsorption of gaseous material from the air by charcoal was
noted by C. W. Scheele in 1777. It seems that most of the people
who have studied adsorption since then have developed formulas
for the adsorption reaction. These formulas are usually called
adsorption isotherms because adsorption is temperature dependent
and the equations must be applied at only one temperature for a
given set of empirical constants--the lower temperatures are more
favorable for adsorption, but the relationship is not simple enough
to include temperature terms in the equation. The most notable
isotherms for engineering purposes are probably those of Langmuir
and Freundlich (or Küster), with theoretical treatments generally
using the isotherms of Gibbs and Brunauer, Emmett, and Teller
(B-E-T).

The Langmuir isotherm, which is based on the theory of the
portion of active sites occupied, is

$$x = fpx_m/(1 - fp),$$ (17-1)

where x = amount of adsorbate, p = pressure of adsorbate in gas
phase, x_m = amount of adsorbate for a monolayer (saturation),
and f = ratio of the rates of adsorption to desorption.

The Freundlich isotherm, an empirical formula, is

$$x/M = kp^{1/n},$$ (17-2)

where x and p are the same as above, M = mass of adsorbent,
and k and n = parameters of fit. n is usually ≥ 1. Of course,
Eqs. 17-1 and 17-2 may use concentrations rather than pressures
since the two are directly related.

EXAMPLE 17-1: How long will it take to saturate a 6 in.-thick
   conical bed of activated carbon by a waste stream with
   64 ppm_v of a solvent that has a molecular weight of 96 and
   follows the Freundlich isotherm such that k = 0.30 atm^{-n}
   and n = 3 if the superficial velocity is 75 ft/min and the
   bulk density of the carbon is 40 lb/ft^3? Assume the temperature is 35°C.
   Adsorptive capacity: $x/M = 0.30(64/10^6)^{1/3} = 0.0120$

   $x/ft^2 = 0.012(40 \text{ lb/ft}^3 \text{ X } 6/12 \text{ ft}) = 0.24 \text{ lb/ft}^2$

   Conc. $= \dfrac{64 \text{ ft}^3}{10^6 \text{ ft}^3} \dfrac{96 \text{ lb/lb mole}}{385(308/293) \text{ ft}^3/\text{lb mole}} = 15.18 \text{ X } 10^{-6} \text{ lb/ft}^3$

   Time to saturation:

   $\dfrac{0.24 \text{ lb/ft}^2}{(15.18 \text{ X } 10^{-6})(75 \text{ ft/min})} = 211 \text{ min}$

Materials transport considerations may be used in the design
of adsorber beds. The transport across the gas stream boundary
layer is as it is for the previously covered absorbers with allow-
ances for the medium; i.e., the rate may be written as

   $r = K_G a \epsilon /\rho (p - p^*),$ (17-3)

where r = adsorption rate, K_G = transfer rate coefficient, a = external surface of adsorbent, ϵ = voids ratio for bed (space among
particles to total ved volume), ρ = bulk density of adsorbent, and
p and p* = adsorbate pressures in the gas phase and at interface,
respectively. The K_G of the equation is sometimes considered to
be composed of two parts: one for the rate of movement through
the boundary layer and the other for the rate of actual adsorption
after reaching the surface. The state-of-the-art for adsorber
design hardly warrants such refinement.

Fair develops a "height of gas transfer unit" relation as

$$-\frac{\partial Y}{\partial Z} = \frac{1}{H_{OG}} \, (Y_A - Y_A^*),$$
(17-4)

where Y_A and Y_A^* = amount of component A (adsorbate) in the gas phase and at the interface, respectively, Z = depth in bed, and H_{OG} = height of gas transfer unit (ft) = $Sc^{2/3}/(aj_d)$, Sc = Schmidt number = $\mu/(\rho_G \mathcal{D})$, a = external suraface (ft^2/ft^3), j_d = Chilton-Colburn mass transfer factor = $1.3(d_p G/\mu)^{-0.45}$, μ = viscosity of gas (lb/ft-hr), ρ_G = gas density (lb/ft^3), $\mathcal{D}$ = diffusivity of gas (ft^2/hr), d_p = particle diameter (ft), and G = mass gas velocity $(lb/hr-ft^2)$.

The amount of heat liberated during adsorption is the heat of liquefaction plus the heat of wetting. The heat liberated may be calculated from two or more isotherms by the Clausius-Clapeyron equation as follows:

$$H = R\frac{d(\ln p^*)}{d(1/T)},$$
(17-5)

where H = amount of heat liberated, R = gas constant, p^* = pressure of adsorbate at interface (equilibrium pressure), and T = absolute temperature.

Despite the number of equations and techniques available for the design of adsorptive air cleaners, most designs are simply scaled up from pilot plant or model data with little or no cognizance of the theoretical parameters. Design factors to be determined are size, configuration, number of units, and manner of operation. These factors depend on the retentivity, contact time (velocity and bed thickness), head loss, heat dissipation, and cycle times.

Activated carbon, the only adsorber that works well in the presence of moisture (the situation in nearly all air cleaning), is made from various materials, including coal, lignite, hardwood, coconut hulls, peach pits, peat, and animal bones. Specifications are often based on the adsorbing capacity (adsorptivity) and the

retention capacity (retentivity). Adsorptivity for specification is
measured by passing a gas stream with carbon tetrachloride in an
amount that would be saturated at $0^{\circ}C$ until there is no further
weight gain for the adsorber. The retentivity is measured by the
amount of the adsorbed material above that will remain after passing
air without any carbon tetrachloride through the bed at $25^{\circ}C$ and
1 atm for 6 hr. Minimum specified values may be $50\%_w$ and $30\%_w$
for the adsorptivity and retentivity, respectively. Other specifications usually list minimum values as follows: apparent density,
25 lb/ft^3; hardness, $80\%_w$ of 6-8 mesh retained on 14 mesh sieve
after shaking 30 min with 30 steel balls (1/4 to 3/8 in. diam.)
per 50 gm of adsorber on a shaker or rapper; material for manufacturing the carbon; and impregnating chemical for reactive adsorption.

Activated carbon usually weighs 35-45 lb/ft^3. Its adsorp-
tivity at $20^{\circ}C$ may range from $15\%_w$ for sulfur dioxide to $50\%_w$ for
acetone to $>100\%_w$ for carbon tetrachloride. The retentivity is
often about one-half the adsorptivity or $5-50\%_w$, depending on the
difficulty of adsorption. The reduced adsorption at normal pressures (concentrations) means that applications may be made at a
few or several $\%_w$ adsorption.

The residence times or contact times vary widely in practice
and range from a few hundredths of a second to several minutes--
the short residence times for recirculating or other applications
where low efficiencies will suffice and the long contact times for
the deep beds used in solvent recovery operations for rather high
concentrations. The residence times depend on the superficial
or approach velocity, the bed thickness, and the volume of voids:

$$\tau = Z\epsilon/v \ , \tag{17-6}$$

where τ = residence or contact time, ϵ = external voids ratio,
v = approach or superficial velocity, and Z = thickness or depth
of bed. The superficial or approach velocity is used because the

actual interstitial velocities are unknown. The usual range for approach velocities is 30-120 ft/min, but the velocities may be much higher for some recirculating, low-efficiency-per-pass applications. The external voids ratio (ε) will run from 0.35 to 0.5. The bed thickness may range from 1/2 in. to several feet, but is usually less than 18 in., and thin beds such as cylinders and accordion beds are often 2 in. thick.

The bed thickness and the head loss are held down by providing a large cross-sectional area of adsorber. In order to obtain a large surface area in a small space, the activated carbon and its support may be put into a conical, cylindrical, or pleated (accordion) configuration (see Fig. 17-1). Whatever the configuration, the exposed holder must be kept full; small bypass areas can give large untreated volumes and thwart the purpose of the adsorber.

The head loss depends on some of the same parameters as the residence time. An equation found by Ergun to represent the head loss is

$$\frac{\Delta P}{Z} = \frac{1}{g} \frac{150(1-\varepsilon)^2}{\varepsilon^3} \frac{\mu v}{d_p^2} + 1.75 \frac{1-\varepsilon}{\varepsilon^3} \frac{\rho_G v^2}{d_p} , \qquad (17\text{-}7)$$

where $\Delta P/Z$ = head loss per foot of depth (lb_f/ft^2-ft), g = gravita-

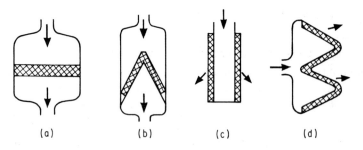

(a)   (b)   (c)   (d)

FIG. 17-1. Configurations for adsorption beds. (a) Flat, (b) conical, (c) cylindrical, and (d) pleated.

tional constant $(lb_m\text{-}ft/lb_f\text{-}sec^2)$, ε = external void fraction of
packed bed, μ = viscosity of gas $(lb_m/ft\text{-}hr)$, v = superficial or
approach gas velocity (ft/sec), d_p = effective particle diameter
in bed (ft), and ρ_G = gas density (lb_m/ft^3).

Perhaps the most used approach to head loss in an adsorber
is a straight-line plot of head loss versus velocity on log x log
paper or

$$h = av^b,   \tag{17-8}$$

where h = head loss (in. w.g./ft of bed), v = superficial gas
velocity (ft/min), and parameters a and b vary with the mesh size
of carbon as follows: (X/Y means passing X mesh and retained on
Y mesh); 2/4, a = $5.07(10^{-4})$ and b = 1.687; 4/6, a = $3.87(10^{-3})$
and b = 1.547; 4/10, 0.00958 and 1.458; 6/16, 0.0322 and 1.334;
10/24, 0.922 and 1.226; 12/30, 0.185 and 1.150.

Heat dissipation is no problem for dilute vapors; for more
concentrated vapors, design may employ dilution and treat larger
volumes of gas or incorporate cooling coils within the adsorber
bed. Surges in concentration can also present problems and must
be considered.

Single adsorption units may be used if the unit can be removed from use long enough to regenerate the activated carbon or
to place in new or renewed units, especially applicable for long
cycle times. Where it is undesirable for the unit to be out of
service during renewal, two or more units may be set in parallel
to permit switching to the clean unit. The adsorption front moves
through the bed with an ogee concentration curve because of the
random nature of the adsorption (see Fig. 17-2). Since break-
through occurs before the bed is saturated, a common design
practice is the use of two beds in series with provision for
switching the direction flow through the beds. Such an arrangement allows saturating a unit before it is regenerated. Three or
four beds may be placed in parallel so that when one is in use,

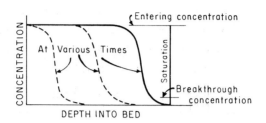

FIG. 17-2. Adsorption front in bed.

one is being regenerated, one is cooling, and the fourth is a
spare.

Regeneration is normally done with steam at less than about
5 psig and saturated, but it may be up to 650°F for application
with high boiling materials in the stream. The adsorbate may not
be removed completely during regeneration because of the cost in
steam for removal of the last fraction. Hot air and/or vacuum can
also regenerate the bed. The amount of heat required during regeneration is that removed during adsorption plus enough to heat
the bed and the bed container to the desired temperature.

IV. APPLICATIONS

Adsorption by activated carbon is widely used in air cleaning
with applications in removing vapors and gases at concentrations
from very dilute to quite large. The adsorption may be on units
which are used once and thrown away (nonregenerative) or on units
which are used for many adsorption-regeneration cycles. The nonregenerative adsorption is not likely to be used on any except very
small concentrations of pollutant.

Physical adsorption cannot be used for true gases, i.e.,
those with critical temperatures <-50°C and boiling points (b.p.'s)
<-150°C. The true gases include nitrogen, oxygen, hydrogen,
carbon monoxide, and methane. Those materials with critical

temperatures in the range of 0°-$150^\circ C$ and b.p.'s -100°-$0^\circ C$ are adsorbed with difficulty after long contact times; the gases diffi- cult to adsorb include hydrogen sulfide, ammonia, hydrogen chloride, formaldehyde, and ethylene. Vapors with b.p.'s $>0^\circ C$ are readily adsorbed.

Activated carbon adsorption is commonly used to collect odors in such places as gymnasiums, nuclear submarines, rendering plants, food processing plants, hospitals, laboratories, printing plants, and perfumeries. Adsorption is the principal removal process in gas masks.

Adsorption is best used in air pollution control for streams with solvent vapors at concentrations less than a few hundred ppm_v. Solvent vapors from degreasing operations offer a classic applica- tion example; the vapors that escape the condensation by the cool- ing coils are removed in carbon beds. In order to use adsorption, the gas stream must be free of particles. For this reason, many otherwise possible applications use incineration or another clean- ing method instead of adsorption. This is especially so where combustible particulates are involved.

V. COSTS

Costs for adsorption pollution control can be described only in the most general, wide-ranging terms or cited for specific installations. The unit costs reported in the literature for adsorp- tion equipment have different bases--per Ncfm, per lb of adsorbent, per lb of adsorbate, or per volume of the process unit.

Based on extrapolations of pilot plant studies, the installed cost of activated carbon adsorbers and auxiliary equipment for collecting protective coating vapors were estimated on a per cfm basis at $6.80-8.00 for a 1000-cfm system, $2.50-3.00 for 10,000 cfm, and $1.54-1.70 for 50,000 cfm. Another estimate that is based on $15\%_w$ adsorption, $0.75 per lb of activated carbon,

and adsorber unit hardware at 5.5 times the carbon cost arrives
at an installed cost of $32.50 per lb of adsorbate capacity.

Some other cost-estimating data that are used include the
following: operating costs of $10 per lb of capacity per year or
0.2-1.0 ¢ per lb of adsorbate removed; regenerating costs, 3-5
lb steam per lb of vapor or dry heat 0.1-0.15 kWh per lb of vapor;
carbon replacement, 1 lb carbon per 2000 lb vapor; cooling water,
7-10 gal per lb of vapor; bed volume of 5.5 ft^3 per 1000 scfm; and
pumping costs of 5 HP per 1000 scfm.

VI. SUMMARY

Adsorption can prevent many air pollution problems by removing polluting vapors and sometimes gases. The adsorbent is
nearly always activated carbon, the only adsorber that works in
the presence of water or its vapor. The waste gas stream is
usually passed through a stationary bed of adsorbent particles.
Nonregenerative or throwaway adsorber units may be used for very
dilute streams where the units will last for a relatively long time,
up to several months; however, the adsorber beds for more concentrated vapors are regenerated and reused continually. The beds
are normally regenerated by steam stripping the adsorbate.

PROBLEMS

1. Plot x/M versus p on log-log paper for Freundlich isotherms
    with n = 1, n = 2, and n = 1/2. Where is k found on such
    a plot?

2. The usual adsorption is assumed to occur at n > 1, which is
    termed a favorable condition for adsorption. Why is this
    description used and why is n < 1 unfavorable?

3. What is the approximate adsorption surface area (acres) for a
    pound of activated carbon?

4. How long will an 18 in.-thick bed last before regeneration if
    the approach velocity is 60 fpm, the concentration of tetra-
    chloroethylene is 20 ppm_v, and the amount collected before
    recycle is 8%$_w$ on a carbon with a density of 40 lb/ft^3 ?

5. Plot $1/x$ versus $1/p$ for Langmuir's isotherm. What is the intercept? What is the slope?

6. From the literature on adsorption, find the six types of adsorption isotherms and briefly describe each type.

7. Plots of the amount of steam (lb of steam/lb of solvent) versus steaming time for a particular adsorption bed show minima at 90 min and 4.1 lb of steam per lb of perchloroethylene and at 75 min and 9.5 lb of steam per lb of toluene. What are the probable reasons for the differences in the two curves?

BIBLIOGRAPHY

Adams, R. E., and W. E. Browning, Jr., Removal of Radioiodine from Air Streams by Activated Carbon, Atomic Energy Commission Rept. ORNL-2872, Washington, D. C., 1960.

Barnebey, H. L., " Removal of Exhaust Odors from Solvent Extraction Operation by Activated Charcoal Adsorption," J. Air Pollution Control Assoc., 15:9, 422, September 1965.

Brunauer, S., The Adsorption of Gases and Vapors, Princeton U. P., Princeton, New Jersey, 1943.

Control Techniques for Hydrocarbon and Organic Solvent Emissions from Stationary Sources, National Air Pollution Control Administration Publ. No. AP-68, Washington, D. C., March 1970.

Danielson, J. A., Ed., Air Pollution Engineering Manual, National Center for Air Pollution Control Publ. No. 999-AP-40, Cincinnati, Ohio, 1967.

Emmett, P., Lecture Notes, Oak Ridge National Laboratory, Summer 1966.

Fair, J. R., "Sorption Processes for Gas Separation," Chem. Engr., 76:15, 90-110, July 14, 1969.

Johnston, W. A., "Designing Fixed-Bed Adsorption Columns," Chem. Engr., 79:26, 87-92, November 27, 1972.

Ross, R. D., Air Pollution and Industry, Van Nostrand Reinhold, New York, 1972.

Strauss, W., Industrial Gas Cleaning, Pergamon Press, New York, 1966.

Young, D. M., and A. D. Crowell, Physical Adsorption of Gases, Butterworth, London, 1962.

Chapter 18

COMBUSTION AND OTHER METHODS

I. INTRODUCTION

Although the air pollutants which are receiving the most
attention are generally derived from combustion processes, combustion is often an effective means of air pollution control.
Combustion can change hazardous or obnoxious materials to
relatively innocuous forms.

Condensation may be used to remove a vapor, or sometimes
a gas, from a waste gas stream. Tank storage and ground injection have been used to provide time for radioactive decay.
Chemical and/or biological conversions appear to be distinct
possibilities for treating some gaseous pollutants.

II. COMBUSTION

Combustion is a rapid oxidation reaction that is initiated
by heat. Fuels give off more heat during combustion than is
required for their combustion.

A. Theory

Combustion results when the combustible molecule contacts the oxygen molecule if the temperature is sufficient to
cause the reaction; i.e., the temperature is above the ignition
temperature. The contact of the two molecules is a probabilistic
event. A finite time is required for the contact to be made; the
required time is shortened by turbulent mixing to increase the
probability of contact. These factors are often cited as the
three T's of combustion--temperature, time, and turbulence.

AIR POLLUTION

There are three distinct types of combustion--flame,
thermal, and catalytic. When the heat content of the gases
being burned lies within the flammable limits, the lower (LFL)
and the upper (UFL) limits, combustion takes place above the
autogenous combustion temperature, resulting in a self-
propagating flame. The flammable limits depend on the conditions;
however, the LFL is roughly 50% of stoichiometric and the UFL
ranges over 165-330% of stoichiometric for the common fuel gases
(C_1-C_4, see Table 18-1).

EXAMPLE 18-1: If the LFL is 52 Btu/Ncf and the gas stream can
be treated as air for specific heat and density, what temperature rise could result from combustion at the LFL?

$c_p \approx 0.24$ Btu/lb-^{O}F and $\rho_a = 0.075$ lb/Ncf   (Fig. 18-1 for c_p)
Possible temperature rise:

52 Btu/Ncf/(0.075 lb/Ncf X 0.24 Btu/lb-^{O}F) = 2890°F

Flame combustion is evidenced by the emission of light
as well as heat. If the combustion air is premixed with the fuel,

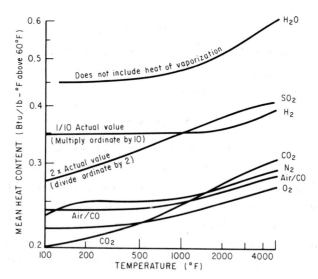

FIG. 18-1. Mean heat capacities at constant pressure.

TABLE 18-1

Combustion Characteristics of Common Combustibles[a]

Substance	Molecular or atomic wt.	Heat of Combustion HHV (Btu/lb)	Heat of Combustion LHV (Btu/lb)	Flammable limits (%v) LFL	Flammable limits (%v) UFL	Combustion with theoretical amount of air Required O2	Combustion with theoretical amount of air Required Air	Products (lb/lb) N2	Products (lb/lb) CO2	Products (lb/lb) H2O	Products (lb/lb) SO2	Products (lb/lb) Total
Carbon (coal)	12.01	14,447	14,447	--	--	2.67	11.50	8.83	3.67	--	--	12.50
Carbon monoxide	28.01	4,344	4,344	12.5	74	0.57	2.46	1.89	1.57	--	--	3.46
Sulfur	32.06	3,980	3,980	--	--	1.00	4.31	3.31	--	--	2.00	5.31
Hydrogen	2.016	60,958	51,571	4.0	75	7.94	34.34	26.40	--	8.98	--	35.38
Hydrogen sulfide	34.08	7,180	6,620	4.3	45	1.41	6.10	4.69	--	0.53	1.88	7.10
Ammonia	17.03	9,668	8,001	--	--	1.41	6.08	5.49	--	1.59	--	7.08
Methane	16.04	23,861	21,502	5.3	14	4.00	17.27	13.27	2.94	2.25	--	18.26
Propane	44.09	21,646	19,929	2.0	12	3.64	15.70	12.06	2.99	1.64	--	16.69
Butane	58.12	21,293	19,665	1.9	8.5	3.58	15.44	11.86	3.03	1.56	--	16.45
Octane vapor	114.23	20,747	19,256	0.8	7.0	3.50	15.10	11.60	3.08	1.42	--	16.10
Ethylene	28.05	21,625	20,276	2.8	29	3.42	14.75	11.33	3.14	1.29	--	15.76
Acetylene	26.04	21,460	20,734	2.5	80	3.07	13.23	10.16	3.38	0.69	--	14.23
Benzene vapor	78.11	18,172	17,446	1.4	7.1	3.07	13.23	10.16	3.38	0.69	--	14.23
Toluene vapor	92.13	18,422	17,601	1.3	6.8	3.13	13.50	10.35	3.35	0.78	--	14.48
Naphthalene vapor	128.16	17,300	16,700	--	--	3.00	12.93	9.93	3.44	0.56	--	13.93
Methyl alcohol vapor	32.04	10,270	9,080	6.7	36	1.60	6.90	5.30	1.37	1.13	--	7.80
Ethyl alcohol vapor	46.07	13,170	11,930	3.3	19	2.08	8.96	6.89	1.91	1.17	--	9.96
Lignite[b]	--	12,055	11,505	--	--	2.14	9.22	7.08	2.71	0.43	--	10.22
Bituminous coal[b]	--	14,550	14,055	--	--	2.60	11.21	8.60	3.08	0.52	--	12.21
Anthracite[b]	--	15,230	14,940	--	--	2.74	11.83	9.08	3.46	0.28	--	12.83

[a]Adapted from Kent, Mechanical Engineers' Handbook on Power, 12th ed., Wiley, New York, 1950.

[b]Computed from moisture- and ash-free averages of 17 lignites, 27 bituminous coals, and 5 anthracite coals.

a hot blue flame is produced; however, if the fuel is burned
without premixing, a yellow flame shows the presence of incandescent carbon from the thermal cracking of the fuel. Sustained
temperatures between those required for ignition and flammability
produce a thermal combustion. This type of incineration requires
longer reaction times than does the flame combustion. Catalytic
combustion permits the flameless combustion of fuel material at
much lower, often $500^{\circ}-800^{\circ}F$ lower, temperatures than thermal
incineration.

B. Design

The different types of incineration necessitate different
equipment designs. Direct flame incineration is carried out by
either a flare or an in-line burner in a furnace; thermal incineration is done in a two-stage furnace, flame combustion and
thermal combustion; and catalytic incineration takes place in a
preheater-catalytic combustion chamber combination. Some
useful combustion design data are shown in Table 18-1 and Fig.
18-1.

1. Direct Flame Incineration

Furnace incineration of pollutants may be of the direct
flame (in-line burner) variety in which all the polluted gas stream
actually passes through the flame front. This type of design is
economical only if the waste gas can be used for the combustion
gas in some boiler or heating facility that is needed anyway.
The burner for direct flame design is generally of the premix type
and provides a velocity pattern such that the flame propagation
velocity is exceeded in the nozzle as a safeguard against flashback problems. The flame velocities for the common gaseous
fuels are 1.1-1.3 fps, but for hydrogen, the velocity is 8.7 fps.

Furthermore, concerns for safety or compliance with in-
surance regulations often limit the heating value of the gases

being carried to the furnace to 25% of the LFL or about 13 Btu/Ncf.
Therefore, in essence, safety precautions require that most of
the fuel be added at the furnace and practically limit the use of
direct flame furnace combustion to boilers. Besides the use of
velocities above the flame propagation velocity and the limiting
of heat values in the incoming gases, other safety procedures
are often employed to prevent explosions. Flame arresters such
as screens or perforated plates that do not get as hot as the
ignition temperature of the gas during operations or water seals
or sprays may be used in connection with the afterburner. For
safe operation there must be provision for automatically shutting
off the fuel in case the flame goes out.

Excess air, an amount of air over and above the stoichiometric requirements, is added in order to burn nearly all the
combustible material within a practicable time; however, the
excesses in the combustion process variables should be used
guardedly because they are not only wasteful of fuel but also
productive of excessive nitric oxide or unburned fuel, pollutants
in themselves. The flame burns much better and is easier to
control when the heat content of the combustion mixture is about
90 Btu/Ncf rather than near the LFL; therefore, calculations of
fuel requirements should be made on this basis.

Gaseous waste streams with high heating values are
usually flared. The gas streams are often put through a dip-leg
water seal to prevent flashback and through a demister to take
out condensed combustibles. Flaring aromatic and multiple-
bonded hydrocarbons (olefins and acetylene) with their low
hydrogen-to-carbon ratios produces large quantities of black
smoke unless the design is made to burn them smokelessly.
The smokeless flare injects live steam through jets surrounding
the flare tip (see Fig. 18-2). The steam induces turbulence into
the periphery of the flame and adds heat to the combustion process,

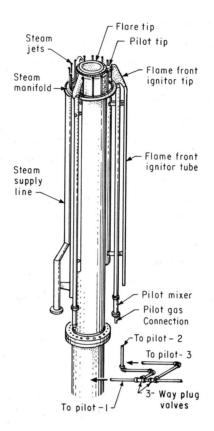

FIG. 18-2. Smokeless flare. From R. J. Ruff,"Nuisance Abatement by Combustion," Air Pollution, Vol. II (A. C. Stern, ed.), Academic Press, New York, 1962.

but perhaps most importantly, the steam produces the water gas reaction with hot carbon according to

$$C + H_2O \longrightarrow CO + H_2 . \qquad (18-1)$$

The carbon monoxide and the hydrogen both burn with blue flames and without smoke. The steam requirements for smokeless incineration range from 0.05 to 0.3 lb of steam per lb of gas, depending on the aromatic and olefinic content of the gas stream.

The flare must be equipped with a pilot light to reignite the gas stream if flameout occurs. The pilot light is usually designed to stay lighted even in high winds (speeds >100 mph).

2. Thermal Incineration

The most widely used furnace afterburner is the thermal incinerator (see Fig. 18-3). Flame combustion is used to produce very hot gases (about 4000°F); then the waste stream is mixed with the hot gases to give temperatures in excess of those required to burn the pollutant (1000°-1500°F). The mixture of gases is kept at this elevated temperature long enough for the combustion to occur. The combustion air for the flame combustion may be a split of the waste gas stream.

EXAMPLE 18-2: What are the fuel requirements for incinerating 1335 acfm of waste gas at 1400°F if the gas has the properties shown? Heating value = 9 Btu/Ncf and Temperature = 150°F.

Assume 50 Ncfm of natural gas will be required.
Combustion gases at 150°F = 665 acfm, total of 2000 acfm.
Composition: 81%$_v$ nitrogen and inerts; 4%$_v$ carbon dioxide; 11%$_v$ oxygen; 4%$_v$ water vapor.

Heat content of waste stream at 150°F, above base 60°F:
Assume composition same as final and use specific heats from Fig. 18-1.

Total flow: 1335 acfm/[385(610/528) ft^3/lb mole]
= 3.00 lb mole/min

		Btu/min
N_2:	0.81 X 3.00 X 7.00 X 90	1531
CO_2:	0.04 X 3.00 X 9.10 X 90	98
O_2:	0.11 X 3.00 X 7.10 X 90	211
H_2O:	0.04 X 3.00 X 8.15 X 90	88
	Total	1928

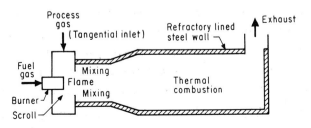

FIG. 18-3. Afterburner of horizontal type.

Note: Calculations with the c_p for air (0.24 Btu/lb-°F) give a value of 1872 Btu/min.

Heat content of exit gases at 1400°F, above 60°F base:

N_2:   0.81 X 4.50 X 7.30 X 1340   35,655
CO_2: 0.04 X 4.50 X 11.5 X 1340   2,774
O_2:   0.11 X 4.50 X 7.70 X 1340   5,107
H_2O: 0.04 X 4.50 X 8.90 X 1340   2,147
                                                    Total   45,683

Additional heat required:

45,683 - 1928 = 43,755 Btu/min + 10% for heat loss
    - heat value of gas = 48,130 - 9(1335)(528/610)
    = 37,700 Btu/min

At this point, the composition could be adjusted for about 40 Ncfm of natural gas rather than the 50 Ncfm used in these calculations and the calculations redone, but such a procedure is probably not warranted.

The proper design of an afterburner for thermal incineration incorporates the three combustion factors, namely, the maintenance of the combustible entities at a sufficient temperature for a period of time that ensures contact of the hot molecules of the pollutant with those of oxygen. The design temperature used is the temperature of the mixture of gases exiting the combustion chamber. The time of contact is calculated as the residence time of the gases in the combustion chamber. The oxygen required for flame combustion plus that required for the thermal combustion must be present either in the waste gas or the added air. The waste gas is introduced in a geometry and at a velocity that will promote turbulence. Usual afterburner designs incorporate gas velocities in the throat of the afterburner of 15-25 fps and in the combustion chamber of 10-15 fps, retention times of 0.3-0.5 sec, and operating temperatures of 1000°-1500°F. The American Petroleum Institute recommends that the maximum velocity in the combustion chamber should be held to 5 fps and that the volume of the combustion chamber should be 1 ft^3 for each 15,000- 20,000 Btu/hr of heat from the fuel and the waste stream.

The temperature which must be sustained for the reaction is the ignition temperature of the pollutant. The flame is caused

by the autogenous combustion temperature having been reached
in the fuel-air mixture.

EXAMPLE 18-3: What should be the approximate dimensions of
a furnace to burn the gas of Example 18-2 by thermal
incineration? Assume plug flows.

Volume of gas at $1400^{\circ}F$:

2000(1860/610) = 6100 acfm = 101.6 acfs

Mixing throat: Use 20 fps and 0.1 sec mixing time.

101.6 acfs/20 fps = 5.08 ft^2 Diameter = 2.54 ft

101.6 acfs X 0.1 sec = 10.16 ft^3
Length - 10.16/5.08 = 2.00 ft

Combustion chamber: Use 15 fps and 0.5 sec retention.

101.6 acfs/15 fps = 6.78 ft^2 Diameter = 2.94 ft

101.6 acfs X 0.5 sec = 50.8 ft^3 Length = 7.49 ft

3. Catalytic Incineration

Catalytic combustion of many air polluting substances can
be carried out at temperatures several hundreds of degrees below
those required for thermal incineration (see Table 18-2). This
results in much less fuel consumption for catalytic incineration
than for any other type. The catalysts used for air pollution
control have generally been either platinum or palladium and
their alloys. Many other metals are being tested in connection
with auto exhausts. The catalyst is supported on porcelain rods,
activated alumina, or ribbons of high-nickel alloy. The catalyst
usually has a surface area of 0.2-0.5 ft^2/Ncfm.

The inlet gas to the catalytic combustor is preheated to a
temperature above that required for initiating the catalytic combustion. This preheating is sometimes done by heat exchange
from the gas stream as it exits from the catalytic unit. The fuel
values for catalytic combustion are nearly always less than 10
Btu/Ncf (temperature rise of $670^{\circ}F$) and, by the use of heat
exchange, even less fuel need be used. Catalytic combustion
requires very little excess air; a design figure of $1\%_v$ is often
used.

TABLE 18-2

Inlet Temperatures for Catalytic Combustion[a]

Industrial process	Contaminating agents in waste gases	Approximate temperature required for catalytic oxidation (°F)
Ammonia manufacturing	Hydrogen sulfide	350–650
Asphalt oxidizing	Aldehydes, anthracenes oil vapors, hydrocarbons	600–700
Carbon black mfg.	Hydrogen, carbon monoxide, methane, carbon	1200–1800[b]
Catalytic cracking units	Carbon monoxide, hydrocarbons	650–800
Coke ovens	Wax, oil vapors	600–700
Formaldehyde mfg.	Hydrogen, methane, carbon monoxide, formaldehyde	650
Nitric acid mfg.	Nitric oxide, nitrogen dioxide	500–1200[c]
Metal lithography ovens	Solvents, resins	500–750
Octyl-phenol mfg.	Phenol	600–800
Phthalic anhydride mfg.	Maleic acid, phthalic acid, naphthaquinones, carbon monoxide, formaldehyde	600–650
Polyethylene mfg.	Hydrocarbons	500–1200
Printing presses	Solvents	600
Varnish cooking	Hydrocarbon vapors	600–700
Wire coating and enameling ovens	Solvents, varnish	600–700

[a]Basic Engineering Principles of the Oxy-Cat, Oxy-Catalyst, Inc., Berwyn, Pennsylvania (now at West Chester, Pennsylvania).

[b]Except carbon 800–1000; temperatures >1200 required to oxidize carbon.

[c]Reducing atmosphere required.

EXAMPLE 18-4: What are the heat requirements for catalytic
    combustion of the gas in Example 18-2 if the characteristics
    are approximately the same as for air and the catalytic
    ignition temperature is $700^\circ F$? What will be the tempera-
    ture of the exit gas without heat exchange?

Heat content of inlet gas, above $60^\circ F$ base:

    3.00 lb mole/min X 7 Btu/lb mole-$^\circ$F X (150° - 60° F)
    = 1890 Btu/min

Heat content to catalyst, from $60^\circ F$ to $700^\circ F$:

    3.00 X 7.2 X (700 - 60) = 13,820 Btu/min

Added heat required:

    13,820 - 1890 = 11,930 + 10% loss = 13,100 Btu/min

Temperature rise in catalyst section:

$$\frac{(9 \text{ Btu/Ncf})/(0.075 \text{ lb/Ncf})}{(7.65 \text{ Btu/lb mole-}^\circ F)/(29.09 \text{ lb/lb mole})} = 456^\circ F$$

Exit temperature:

    700 + 456 = $1156^\circ F$

Note that 60% heat recovery would require a heat content of
    13,820(0.40)/(1335 X 528/610) = 4.78 Btu/Ncf in order
    to use catalytic combustion without additional fuel.

C. Applications

Combustion can be used to control many air pollution
problems. These applications include the destruction of odors,
toxic substances and infective agents, smoke and opacity-
producing materials, and reactive materials, the prevention of
explosion hazards, and the reduction of pollutants in the oxidized
forms. The many and diverse process applications of incineration for
air pollution control include aluminum chip driers, animal blood
driers, asphalt-blowing stills, automotive brake shoe debonding
ovens, citrus pulp driers, coffee roasters, electric insulation
burnoff ovens, flue-fed refuse incinerators, foundry core ovens,
meat smokehouses, paint-baking ovens, rendering cookers,
rubber curing, varnish cookers, wire enameling ovens, paper
printing and impregnating, pharmaceutical manufacturing, pesticides manufacturing and blending, and textile finishing.

Waste gas streams with high fuel values, above the explosive limit, are likely to be flared without any attempt at heat
recovery. This is especially true for varying compositions and
flow rates.

Direct flame (in-line) combustion of air pollutants is limited
to those installations where a boiler or other heater application
can make use of the heat generated by the combustion or where
the flow rates are small enough to keep the fuel costs reasonable.

Thermal incineration is carried out in most afterburners
where the waste stream does not contain significant fuel values.
This type of incineration is most popular for relatively low air
flows because the fuel requirements are considerable.

Catalytic combustion requires that the gas stream be relatively clean of particulates and certainly free of metallic substances which could plate out on the catalyst and poison it.
The catalyst bed may be cleaned on a periodic basis (quarterly
or annually) by scrubbing with water and sometimes acid and
often by heating to more than $1100^{\circ}F$ in those applications where
the operating temperatures are much lower than $1100^{\circ}F$. Reduction of an oxidized pollutant such as nitrogen dioxide or nitric
acid is usually done by using the material as the oxygen source
for some added fuel in catalytic combustion.

D. Costs

The costs for incineration vary widely with the individual
application, as do all air pollution control processes. The cost
for the basic equipment, without installation, runs from about
$1 to $3 per Ncfm for relatively large afterburners (10,000-20,000
Ncfm). Catalytic afterburners are generally the most costly
among the types used and are near the top of the cited cost range.
The cost of the catalyst itself may well exceed $1 per Ncfm.
Maintenance costs of $0.06 and $0.20 per Ncfm have been cited
for the usual thermal afterburners and catalytic afterburners,

respectively. The largest part of the overall annual costs for
incineration is usually the fuel cost. Smokeless flares have
large fuel costs for the steam despite the fact that the gas stream
may be far above the flammable limit in heat content. Fuel costs
vary with the heat content of the waste stream and with the method
of incineration. They are not difficult to calculate and should be
carefully determined.

III. OTHER METHODS

Many methods other than absorption, adsorption, and
combustion are sometimes used for the control of gaseous air
pollutants. These methods include condensation, oxidation
other than combustion, reduction of oxidized pollutants, masking
and counteracting odors, and the use of biological controls.

Condensation of air pollutants may be the principal mechanism in absorption and adsorption processes, but it is sometimes
carried out as a separately designed process. Economic considerations usually limit the application of condensation to vapors, such
as the perchloroethylene vapors from degreasers. However, in
some special instances, actual gases are condensed. One such
application is for the radionuclide krypton-85.

Some air pollutants may be oxidized by means other than
combustion. These oxidations are often done with ozone, potassium permanganate, or chlorine and its compounds. For
example, ozone has been used to oxidize the hydrogen sulfide
in sewer gas at sanitary sewer manholes and lift stations, and
potassium permanganate has been applied to cattle feed lots to
abate the urine and manure odors.

Radioactive gases are stored in tanks and in porous strata
of the earth for several hours up to many years to permit their
racioactive decay before they are released to the environment.

Masking and counteraction of odors have been rather widely

used. Large open processes (such as sewage treatment plants
and oil-producing fields) and closed processes which release
large volumes of gases at many points are not amenable to odor
control by the positive processes such as adsorption or combustion. Masking agents or perfumes are often used to prevent com-
plaints from residents of areas close to the processes. Sometimes actual counteraction of odors can be achieved, but usually
the less desirable masking of odors is the method applied.

Biological control of air pollutants has been used with
green belts between industrial and residential areas. The practice has been more common in Europe than in the United States.
Recent studies have shown soil microorganisms to be responsible
for destroying carbon monoxide. This and other biological treatments of air pollutants may find future application.

IV. SUMMARY

Combustion may be used for air pollution control. The
process consists of maintaining the temperature of combustion
for a sufficient time to insure reaction, with or without a catalyst.
It has been applied to the destruction of process odors, toxic
and infective agents, combustible particles, and reactive, smog-
producing materials. Combustion can be used to control explosion hazards and to reduce oxidized pollutants.

Control of gaseous pollutants may be done by methods
other than absorption, adsorption, and combustion. These other
methods include condensation, oxidation other than combustion,
storage for decay, masking and counteraction of odors, and
biological removal or conversion of a pollutant.

PROBLEMS

1. What is the indicated temperature rise for burning a gas
   stream (assume air) with 90 Btu/Ncf?
2. What will be the approximate actual flame temperature in
   Problem 1? Why?

3. Determine the fuel requirements for Example 18-2 using direct flame combustion.

4. Find the required fuel for Example 18-2 using catalytic combustion with an initiation temperature of 850°F and no heat recovery.

5. Could the catalytic combustion in Problem 4 be sustained without the addition of fuel?

6. Discuss the probable future of control by incineration as affected by increasing fuel costs and stiffening air pollution regulations.

7. Describe the type of condenser that would remove krypton-85 from an air stream. What are the design parameters? boiling point = -152.9°C

BIBLIOGRAPHY

Air Pollution Manual, Part II: Control Equipment, American Industrial Hygiene Assoc., Detroit, Michigan, 1968.

Control Techniques for Hydrocarbon and Organic Solvent Emissions from Stationary Sources, National Air Pollution Control Administration Publ. No. AP-68, Washington, D. C., March 1970.

Danielson, J. A., Ed., Air Pollution Engineering Manual, U. S. Public Health Service Publ. No. 999-AP-40, Cincinnati, Ohio, 1967.

Perry, J. H., R. H. Perry, C. H. Chilton, and S. D. Kirkpatrick, Eds., Perry's Chemical Engineers' Handbook, 4th ed., McGraw-Hill, New York, 1963.

Ross, R. D., Ed., Air Pollution and Industry, Van Nostrand Reinhold, New York, 1972.

Strauss, W., Industrial Gas Cleaning, Pergamon, New York, 1966.

Section IV: Particulate Control

The removal of particulates from a gas stream may be accomplished by mechanical or inertial processes, filtration, scrubbing,
electrostatic precipitation, or combinations of processes. For
most practical applications, there are only three types of air
cleaners capable of meeting the rigorous demands of current air
pollution control regulations--bag filters, high energy scrubbers,
and electrostatic precipitators. Comparative costs for obtaining
various efficiencies of collection by different types of collectors
are presented in Fig. IV, and Table IV shows the relative distribution of the costs of particulate collection among the various cost
factors.

TABLE IV

Approximate Relative Costs (1963) of Various 60,000-cfm Particulate Collectors [68°F (20°C)][a]

Equipment (Key No.)	Effic. on std.[1] (%$_w$)	Capital cost[2] ($)	Cleaner volume (ft³)	Press. drop (in. w.g.)	Power cost[3] ($/yr)	Water usage (gal/ 10³ft³)[4]	Water cost[5] ($/yr)	Maintenance[6] ($/yr)	Running cost[7] ($/yr)	Total cost[8] ($/yr)	Total cost[8] (¢/10³ft³)	Overall equiv. press.[9] (in. w.g.)
Cyclones (2)	65.3	18,000	6,000	3.7	4,060	--	--	144	4,200	6,000	0.021	5.3
Cyclones (4)	84.2	34,600	12,000	4.9	5,420	--	--	144	5,570	9,020	0.031	8.0
Multiple cyclones	93.8	37,700	4,200	4.3	4,750	--	--	144	4,900	8,660	0.030	7.7
Irrigated cyclones	91.0	42,700	9,000	3.9	4,870	4.0	1580	360	6,820	11,090	0.039	9.8
Cyclones (3)	74.2	30,700	3,600	1.4	1,570	--	--	144	1,720	4,790	0.017	4.2
Electrostatic precipitators	94.1	168,000	30,000	0.6	1,490	--	--	600	2,090	18,900	0.066	16.8
Electrostatic precip. (11)	99.0	216,000	36,000	0.6	2,690	2.5	1060	960	4,700	26,300	0.092	23.4
Low velocity fabric filters	>99.9	45,600	42,000	2.0	2,620	--	--	7,820	10,400	15,000	0.052	13.3
Shaker-type fabric filters	>99.9	144,000	60,000	3.0	4,220	--	--	9,120	13,300	27,700	0.096	24.6
Reverse jet fabric filters	99.8	158,000	36,000	4.0	9,430	--	--	14,700	24,200	40,000	0.139	35.5
Spray tower	96.3	100,000	15,000	1.4	5,700	18.0	7920	720	14,300	24,300	0.085	21.6
Scrubber (5)	97.9	56,400	6,000	6.1	6,960	3.0	1320	720	9,000	14,600	0.051	13.0
Scrubber (6)	93.5	47,500	9,000	6.1	6,770	0.6	264	480	7,500	12,300	0.043	10.9
Venturi scrubber (14)	99.7	82,000	9,000	22.0	25,400	7.0	2900	720	29,000	37,200	0.129	33.0
Disintegrator	98.5	97,700	4,200	--	54,500	5.0	2040	480	57,000	66,800	0.232	59.2

152

[a]From Stairmand, see Fig. IV.

[1]$80\%_w < 60$ μm; $45\%_w < 20$ μm; $30\%_w < 10$ μm; $20\%_w < 5$ μm; $12\%_w < 2.5$ μm. Other sizes specified in reference.

[2]Total cost including installation and auxiliaries (but not solids disposal equipment). Money from pounds to dollars.

[3]Cost of electrical energy assumed 1.2¢/kWh with fan and motor efficiency taken as 60%.

[4]Includes power consumed by auxiliaries (but not solids disposal equipment).

[5]Cost of water assumed 0.34¢/m^3.

[6]Estimated figure, including replacement bags for fabric filters.

[7]Assuming 8000-hr operation per annum.

[8]Capital charges taken as 10% of capital cost.

[9]Total annual cost expressed as power usage.

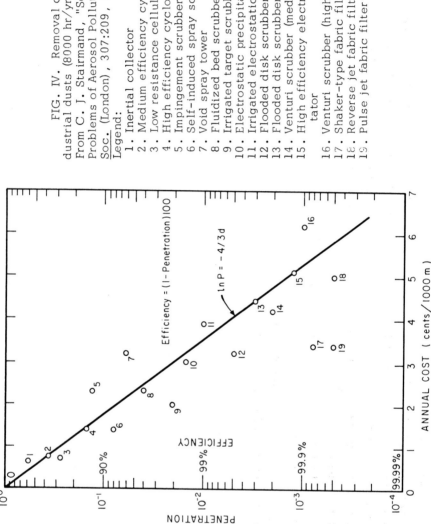

FIG. IV. Removal costs for fine industrial dusts (8000 hr/yr basis--1969). From C. J. Stairmand, "Some Industrial Problems of Aerosol Pollution," Proc. Roy. Soc. (London), 307:209, 1968.

Legend:
1. Inertial collector
2. Medium efficiency cyclone
3. Low resistance cellular cyclone
4. High efficiency cyclone
5. Impingement scrubber (Doyle type)
6. Self-induced spray scrubber
7. Void spray tower
8. Fluidized bed scrubber
9. Irrigated target scrubber (Peabody type)
10. Electrostatic precipitator
11. Irrigated electrostatic precipitator
12. Flooded disk scrubber (low energy)
13. Flooded disk scrubber (medium energy)
14. Venturi scrubber (medium energy)
15. High efficiency electrostatic precipi-
    tator
16. Venturi scrubber (high energy)
17. Shaker-type fabric filter
18. Reverse jet fabric filter
19. Pulse jet fabric filter

Chapter 19

GRAVITATIONAL, CENTRIFUGAL, AND INERTIAL COLLECTION

I. INTRODUCTION

The particle collectors described in this chapter are those devices which separate the particles from the carrying gas stream by gravity, centrifugal force, and inertia. Such devices include settling chambers, cyclones, dynamic precipitators, baffle collectors, and louver concentrators.

These types of cleaners will not generally achieve high efficiency removal of small particles. For practical purposes their applications are limited to removal of particles with diameters in excess of a few micrometers; therefore, the role that gravitational, centrifugal, and inertial collectors can play in air pollution control today is mostly that of precleaner followed by a high efficiency collector or the removal of fine particles after the particles are entrapped in large liquid droplets or have undergone agglomeration.

II. GRAVITATIONAL COLLECTION

Particles may be removed from a gas stream by allowing gravitational settling. If the gas stream with its entrained particulates is slowed sufficiently by an enlargement in the duct, the large particles will settle out. The theory of sedimentation is simplified for equation by assuming spherical particles and using straight-line regressions for three sections of the log plot of the drag coefficient versus Reynolds number (see Chapter 5).

A. Design Concepts

The principal design factors for gravitational particulate

collection are the velocity of settling and the maximum distance that a particle must settle for collection. The residence time required for collecting a given size particle at a certain efficiency provides a parameter which is directly related to the physical size (volume) of the settling chamber.

Laminar flow through the settling chamber is often assumed; then a correction is made for the fact that laminar flow seldom, if ever, prevails in practice where the transport velocities are usually 1-10 fps. Commonly, the design is based on laminar flow, but the theoretical settling velocity for the particles is halved. The equations for settling velocities of spherical particles are divided into the drag regions as follows:

Stokes or Streamline Region--$N_{Re} < 3$:

$$u_t = \rho_p g d_p^2/(18\mu) \text{ or in air @ } 20^\circ C, \; u_t = 0.003 \rho_p d_\mu^2, \quad (19\text{-}1)$$

where u_t = terminal settling velocity (cm/sec), ρ_p = density of particle (gm/cm^3), d = particle diameter (d_p in cm and d_μ in µm), μ = dynamic viscosity of gas (poises or gm/cm-sec), and g = acceleration of gravity ($\sim$981 cm/sec^2).

Intermediate Region--$3 \leq N_{Re} \leq 1000$:

$$u_t = 0.20 (\rho_p g)^{2/3} d_p/(\rho_g \mu)^{1/3}, \quad (19\text{-}2)$$

where ρ_g = density of gas (gm/cm^3).

Turbulent Region--$N_{Re} > 1000$:

$$u_t = (3 \rho_p g d_p/\rho_g)^{1/2}. \quad (19\text{-}3)$$

For the aerosols most often encountered in air pollution control, ρ_p is about 2.7 gm/cm^3 and the particle size covered by Eq. 19-1 goes up to about 83 µm in air at NTP. These equations neglect buoyancy; to include buoyancy replace the density of the particle with the density of the particle minus the density of the gas.

For laminar flow in a settling chamber with dimensions L X W X H, the smallest particle size removed with 100% efficiency

(d_{100}) settles a distance H with a velocity u_t while being transported a distance L with a velocity of v; i.e., the residence time
(t), where t = L/v, is sufficient for the particle with velocity u_t
to settle a distance H:

$$H/u_t = L/v. \qquad (19\text{-}4)$$

If Eqs. 19-1 and 19-4 are combined, the result is

$$d_{100} = [18\mu vH/(g\rho_p L)]^{1/2}. \qquad (19\text{-}5)$$

A frequently used sedimentation design concept is that of overflow
rate (Q/LW), where Q = volumetric flow rate. Since v = Q/WH,

$$d_{100} = [(18\mu/g\rho_p)(Q/LW)]^{1/2}. \qquad (19\text{-}6)$$

Smaller particle sizes should be collected with an efficiency that
is simply equal to the ratio of their settling velocities to that for
the d_{100} particle size; i.e.,

$$\eta_i = 100\,(d_i/d_{100})^2. \qquad (19\text{-}7)$$

This idea may also be stated as the ratio of the distance that the
particles settle while passing through the chamber to the distance
that they must settle for total collection (H).

The flow in a settling chamber will probably be turbulent
rather than laminar. For turbulent flow, the penetration of particles is in accordance with

$$q = \exp(-u_t t/H) = \exp(-N_t), \qquad (19\text{-}8)$$

where q = fraction penetrating. When $u_t t/H = 1$, 37% of the
particles remain suspended in the gas stream. This procedure
may be used instead of doubling the size of the sedimentation
chamber by halving the settling velocity.

EXAMPLE 19-1: What portion of 50μm-diameter particles with
ρ_p = 2.0 gm/cm^3 will be collected with a settling height
(H) of 4 ft and a residence time of 10 sec if laminar flow
prevails? If the flow regime is turbulent?

$$u_t = 0.003\,(2.0)\,(50)^2 = 15 \text{ cm/sec}$$

Distance settled in 10 sec = 15(10) = 150 cm = 5 ft
Since 5 ft is greater than H, 100% collection is made.
Turbulent:

$q = \exp(-15 \times 10/122) = 0.292$

$\eta = 1 - q = 1.00 - 0.29 = 0.71$ or 71% collection

Obviously, the desired configuration for a settling chamber
is the one with a minimum overflow rate, maximum L X W. The
Howard settling chamber (see Fig. 19-1) achieves this end by
using multiple shelves in a chamber. There are problems involved
in keeping the shelves from warping and in removing the collected
materials from the shelves.

Good settling chamber design requires uniform distribution
of the transport velocity over the settling height and width,
storage of the collected material without reentrainment, and provision for the removal of the collected material.

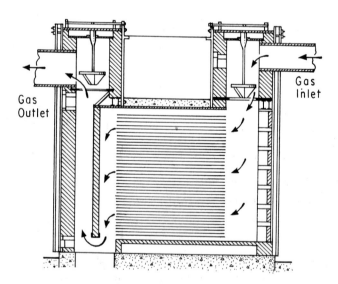

FIG. 19-1. Howard settling chamber. From C. E. Lapple,
"Dust and Mist Collection," Perry's Chemical Engineers' Hand-
book (J. H. Perry et al., eds.), 4th ed., McGraw-Hill, New
York, 1963.

B. Applications and Costs

Settling chambers might find application in the removal of rather large particles (usually >50 μm) from a gas stream which has other cleaning following. They have been used for catching the particles that settled out of plumes immediately adjacent to the stacks on cupolas, kilns, sinter plants, and scarfing machines and for sawdust, wood chips, alfalfa straw, and cotton gin wastes. To build a settling chamber for collecting small particles would be prohibitively expensive; however, a lime plant in Pennsylvania uses an abandoned coal mine as a secondary cleaner after its wet scrubbers.

A first estimate of the purchase cost for settling chambers may be made as $0.20/acfm; the maintenance costs should be practically nil; and the operating costs should be quite low since the principal head losses are those at the inlet and the outlet of the unit.

III. CENTRIFUGAL COLLECTION

Centrifugal collectors put the airstream through a curving path in order to achieve large separation forces on the particles. This group of collectors includes cyclones, multiple cyclones, and centrifuges such as the dynamic precipitator.

A. Theory

The centrifugal force on an object moving in a circular path is calculated by

$$s.f. = F/W = v^2/(gR), \qquad\qquad (19-9)$$

where s.f. = separation factor (gravities, g), F = centrifugal force, W = weight of particle, v = tangential velocity of particle, g = acceleration of gravity, and R = radius of particle path.

B. Design

If the migration velocity (ω) is assumed to equal the separation factor times Stokes settling velocity (a reasonable assumption for a Reynolds number less than 3), the diameter of the particle for 100% collection (d_{100}) will be

$$d_{100} = [18\mu \; \ell n(R_2/R_1)/(\rho_p \omega_a^2 t)]^{1/2}, \tag{19-10}$$

where ω_a = angular velocity (radians/sec), t = residence time or time for particle to move from R_1 to R_2 (sec), R_2 = outer radius of air path, and R_1 = inner radius of air path (ft or cm). The difficulty in applying Eq. 19-10 or a similar equation is that the residence time t is not readily determinable since the flow pattern is not defined.

EXAMPLE 19-2: What is the calculated residence time in 6 in.-diameter multiple cyclone tubes for collecting 100% of 5 μm-diameter ferric oxide particles $(\rho_p = 5.24 \; gm/cm^3)$ if the tangential inlet velocity to the tubes is 70 fps through 1 in.-wide inlets ?

ω_a = 70 fps/0.25 ft = 280 radians/sec

$t = [18(1.8 \times 10^{-4}) \ell n(3/2)]/[5.24(280)^2(5 \times 10^{-4})^2]$

  = 0.0128 sec

Check drag coefficient region:

  Migration velocity = s.f. X u_t

  Highest velocity = $\{(70)^2/[32.2(2/12)]\}[0.003(5.24)(5)^2]$

    = 358 cm/sec

  $N_{Re} = \rho_g vd/\mu = 0.0012(358)(5 \times 10^{-4})/(1.8 \times 10^{-4})$
    = 1.2

  N_{Re} < 3   Equation 19-9 applies.

Another common form for the d_{100} calculation is obtained by assuming that the particle must move through a maximum distance of the inlet width (b) and that

$$t = 2 \, N\pi R/v \tag{19-11}$$

where N = number of turns that the air makes in the collector and v = inlet velocity to the collector; the equation resulting from such assumptions is

$$d_{100} = [9\mu b/(\pi N v \rho_p)]^{1/2}. \tag{19-12}$$

The parameter N is calculated for the cyclone from its performance; its usual range is from 5 to 10. The 50% cut size (d_{50}) may be calculated by using half the migration distance for collection (b/2) or by simply multiplying d_{100} times $1/\sqrt{2}$.

Strauss uses a much more theoretical approach to derive an expression for d_{100}. He also describes eddy currents and pressures in cyclones and gives a rather thorough treatment of the subject of cyclones.

The head loss in a cyclone may be calculated from any of several formulas. One often used formula comes from First; it is

$$\Delta P = \Delta h/(0.003 \rho_g v^2) = \frac{12\,bh/(kp^2)}{(L/D)^{1/3}(H/D)^{1/3}}, \tag{19-13}$$

where ΔP = head loss (velocity heads), Δh = head loss (in. w.g.), ρ_g = density of gas (lb/ft^3), v = inlet gas velocity (ft/sec), b = width of entry (ft), h = height of entry (ft), p = diameter of exit (ft), L = length of cylinder (ft), H = height of cone (ft), D = diameter of cylinder (ft), and k = entry vane factor = 1/2 for no entry vane = 1 for vane which splits annular section = 2 for entry vane that extends to exit duct wall. A head loss formula from Lapple is

$$\Delta P = Kbh/D^2, \tag{19-14}$$

where K averages 13 and ranges from 7.5 to 18.4. The average value gives ΔP within 30% of most experimental data.

Dynamic precipitators are centrifugal pumps that are used as centrifuges to separate particles from the airstream; they are formed to have a channel for collecting particles which come out radially as the airstream is taken off at an angle to the impeller.

Calculations should be made on the basis of the time that the air resides in the centrifugal field, the gravities put on the particles, and the distance particles move for collection.

C. Applications

Far more cyclones have been used for gas cleaning than all other types of control units combined. Few centrifugal collectors other than the cyclone have been used.

Many types of cyclones have been devised (see Fig. 19-2). The most popular types are the conventional cyclone and the multiple cyclone. The high efficiency cyclone is not really a high efficiency collector when compared with other collectors, just when the comparison is with other cyclones. Standard cyclones can effectively collect particles down to about 20 μm, high efficiency cyclones to about 8 or 10 μm, and multiple cyclones to near 4 or 5 μm.

Cyclones have been used extensively not only in collecting dusts that emanated from various processes but also for collecting the airborne product in driers, pneumatic conveyors, and air-sorting crushers. Cyclones have been used singly and in series, the latter being for additional recovery of valuable material such as a petroleum refinery catalyst or detergent or soap powder. In the past, nearly all feed mills (grinding grain, alfalfa, etc.) have had cyclones to recover the air-carried materials. Cyclones have been used for sawdust, wood chips, fertilizer, hot-mix asphalt plants, grinding of metals, crushing and sorting lime, and as precleaners on fly ash. The list of applications is virtually endless because cyclones have been tried almost everywhere that dusts are emitted. Their role at present is solely as a precleaner to some more efficient process that can remove the small particles or following a high-energy scrubber to collect the large liquid droplets and the entrapped small particles, the only exception being the few applications where the multiple cyclone can meet

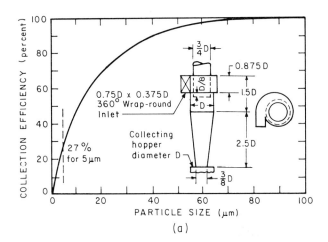

(a)

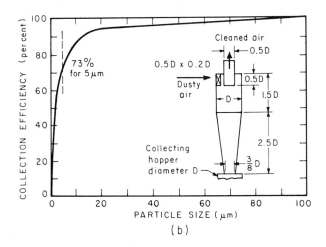

(b)

FIG. 19-2. Cyclone proportions and efficiencies. (a) Stand-ard and (b) high efficiency. From C. J. Stairmand, <u>Gas Purifica-tion Processes</u> (G. Nonhebel, ed.), George Newnes, London, 1964. [In the 2nd ed., <u>Processes for Air Pollution Control</u>, CRC Press, Cleveland, Ohio, 1972.]

the regulations. Cyclones have been irrigated to improve their efficiencies. The wet-type centrifugal collectors are discussed in Chapter 21.

Dynamic precipitators are not widely used. They have the advantage of negative head loss or pumping capability, but they are expensive when designed to collect small particles, and the erosion of the impeller causes high maintenance costs (see Sec. IV, p. 152, Disintegrator).

D. Costs

Although cyclones are relatively low in cost, they are not economical unless they are suitable for the application. Many companies have cyclones lying around their junk piles because they were put onto rather than designed for the process. The purchase cost for cyclones is usually in the range of 20-50 ¢/acfm. For more on centrifugal collection costs, see Sec. IV, pp. 152-154.

IV. INERTIAL COLLECTION

An aerosol stream may be put through a baffled pathway such that inertial impaction will remove the entrained particles. The inertial collection may be the result of turning the gas stream through a rather large angle, up to 180°, or a fairly small deflection as by a fiber. The latter case is described in Chapter 20.

The baffle cleaner has the advantage of simplicity in design and construction (see Fig. 19-3). The major drawback to such a cleaner, aside from low efficiency on small particles, is the difficulty of removing the collected particles from the baffles without reentrainment. Collected particles must be stored in the areas where removal occurs until the buildup sloughs off into the hopper. The principal applications of such cleaners is for the removal of droplets because the collected liquid will run off the baffles.

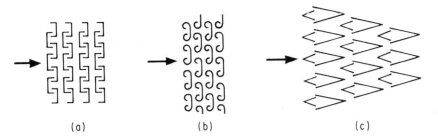

FIG. 19-3. Some types of baffle cleaners. (a) Channel baffle, (b) DEP curtain, and (c) venturi baffle.

The louver separator has been applied to the concentration of particles into a fraction of the total flow. The aerosol stream after concentration by the louver is then put through a cyclone for the particle removal. Its advantage lies in the reduction of the size of the cleaner and the slight benefits gained in treating a concentrated gas stream by a cyclone.

V. SUMMARY

Settling chambers can achieve practical removal only for quite large particles. Some particles which are too small for gravity sedimentation may be separated by centrifugal force in cyclones and dynamic precipitators. The cyclone has been widely used throughout air cleaning for removing medium to large sized particles. Baffle or inertial collectors are sometimes used to collect dusts, but more often they collect mists because the liquid will remove itself from the tortuous shapes of the baffles by runoff, whereas dust removal from these surfaces may be quite difficult.

These types of cleaners have found wide application in the past because they are relatively inexpensive. However, the current rigid air pollution control regulations have relegated gravitational, centrifugal, and inertial collection to the role of precleaner in front of a more efficient process or to the removal of large droplets and agglomerated particles after high-energy scrubbing and/or agglomeration.

PROBLEMS

1. Determine the dimensions of a settling chamber to collect a
   sawdust ($\rho = 0.9$ gm/cm^3) with d_{100} equal 1 mm (1000 μm)
   from a gas stream of 800 acfm at 100°F. Assume that it is
   desirable to have the L·W·H sizes in the ratio of 10:1.2:1.

2. What is the efficiency of collection of the settling chamber in
   Problem 1 for the particles with diameters of 200 μm?

3. What would the migration velocity (ω) be for the particles in
   Problem 2 if the gas stream were put into a 3 ft-diameter
   cyclone at 70 ft/sec?

4. What would the d_{100} size in Problem 3 be if the width of the
   inlet is one-half the height of the inlet and the residence
   time is 0.7 sec?

5. Work Problem 4 with the assumption that the air makes 6 turns
   in the cyclone.

6. What is the d_{50} size for Problem 4?

BIBLIOGRAPHY

Control Techniques for Particulate Air Pollutants, National Air
Pollution Control Administration Publ. No. AP-51, Washington, D. C., January 1969.

Danielson, J. A., Ed., Air Pollution Engineering Manual, National
Center for Air Pollution Control Publ. No. 999-AP-40, Cincinnati, Ohio, 1967.

Drinker, P., and T. Hatch, Industrial Dust, 2nd ed., McGraw-
Hill, New York, 1954.

First, M. W., "Stack Gas Cleaning," Amer. Industrial Hyg. Qtrly.,
11:4, 209, 1950.

Friedlander, S. K., L. Silverman, P. Drinker, and M. W. First,
Handbook on Air Cleaning: Particulate Removal, U. S. Atomic
Energy Commission, Washington, D. C., September 1952.

Lapple, C. E., "Dust and Mist Collection," Perry's Chemical
Engineers' Handbook (J. H. Perry et al., eds.), 4th ed.,
McGraw-Hill, New York, 1963.

Stairmand, C. J., "The Design and Performance of Modern Gas-
Cleaning Equipment," J. Inst. Fuels (London), 29, 58-76,
February 1956.

Strauss, W., Industrial Gas Cleaning, Pergamon Press, London,
1966.

Chapter 20

FILTRATION

I. INTRODUCTION

Filters of many diverse types find application in some phase of air pollution control; however, fabric or bag filters are the usual ones for industrial installations. Some type of filter will meet the efficiency requirements for any gas cleaning; therefore, the applicability of filters depends on their economic feasibility, a factor that is strongly influenced by bag life. Other types of filters that are sometimes applied to air cleaning for pollution control include granular bed, fiber mat or bed, viscous, and absolute or HEPA filters.

II. THEORY

Filters are fabrics or mats or other porous matrices that are intended to pass the gaseous portion while retaining the particles from an aerosol stream. Several mechanisms are involved in the separation of the particles, the principal ones usually being impaction, interception, and diffusional impaction (see Fig. 20-1). The other collection mechanisms include electrostatic forces, gravity, straining (double interception), and movements of particles by radiation, light, heat, and vapor fluxes. The relative influences of some of the latter are negligible and for some are ill-defined.

A. Impaction

Impaction is the mechanism operative in the removal of particles that, because of their inertia as they are moved with the gas stream, cross the streamlines of the gas flow around an

167

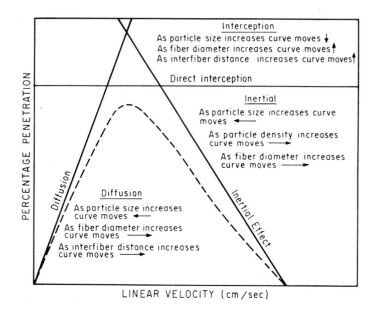

FIG. 20-1. Mechanical mechanisms of fiber filtration.
From E. A. Ramskill and W. L. Anderson, J. Colloid Sci., 6,
416, 1951.

obstacle and impact on the obstacle (see Fig. 20-2 and Chapter 5).
In the normal industrial bag-filter operation, velocities are very
low and impaction plays a relatively minor role; however, the
principal removal action of glass fiber mat filters used in air conditioning is impaction.

B. Interception

Interception is the term applied to the collection mechanism
that removes particles through a surface collision of the particles
and the media as a result of the physical sizes of the particles.
The dividing criterion between impaction and interception is
whether the particle would be collected if it were a point mass;
i.e., if the center of the particle follows a trajectory which would
clear the obstacle but the outer periphery does not, then the removal is by interception, whereas if the center of the particle moves

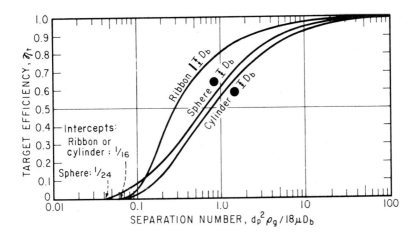

FIG. 20-2. Target efficiency versus impaction parameter.
From S. K. Friedlander et al., Handbook on Air Cleaning: Particu-
late Removal, U. S. Atomic Energy Commission, Washington, D.
C., 1952.

along a path that intersects the surface of the obstacle, the removal is by impaction. Interception is quite important in filtration
and its importance increases as the dust cake forms.

C. Diffusional Impaction

Diffusional impaction results from the Brownian movement of
the particles across streamlines. Since Brownian motion of particles with diameters larger than about a micrometer is nearly nil,
diffusional impaction removes only the small particles. Diffusion
filters are thick and have low velocities through them in order to
give enough time for diffusion to deposit the particles on the filter
media.

III. DESIGN

The first order of business in the design of air pollution control by filtration is the selection of the filter type suited to the
job. Selection is made on the basis of efficiency required, particle size and abrasiveness, grain loading, and temperature. The

decision is made as to whether a fabric, granular, mat, or HEPA
is desirable.

A. Design Parameters

After the decision that a certain type of filter should be used,
or concurrent with and part of that decision, the type of filter
material needs to be selected. This selection is based on economically achieving the objective of cleaning the gas stream to
or beyond the required efficiency. The economics of filtration
involve equipment costs, head loss, and durability or lifetime
of the filter material.

1. Filter Materials

The proper material for the filter matrix is highly important
to the success of filtration. The initial material selection is
most often based on what others are using for similar installations.
The majority of plants using filtration try various filter fabrics
in order to find the most suitable materials available.

a. Fabric. The choice of fabric fiber is based primarily on
operating temperature and the corrosiveness or abrasiveness of
the particles. Table 20-1 lists some of the typical properties for
the most popular fabrics.

EXAMPLE 20-1: What is the packing density (β) for 16-oz Nomex
felt if the fabric thickness is 1/16 in.?

s.g. = 1.38   (from Table 20-1)

$$\beta = \frac{16 \text{ oz/yd}^2}{(1/16 \times 1/12 \text{ ft})(9 \text{ ft}^2/\text{yd}^2)(1.38 \times 62.4 \times 16 \text{ oz/ft}^3)}$$

$$= 0.248$$

The weave and the finish of the fabric affect the efficiency
obtained, the ease of cleaning the fabric, and the life of the
filter material. Because of the cost of and inconvenience in
changing the bags, an overriding criterion in fabric selection,
especially fabric finish, is the frequency of cleaning that the

TABLE 20-1

Typical Properties of Some Filter Fabrics

Material	Max. °F contin. (brief)	Sp. gr.	Chemical resistance[a]				Type (weave ct. or felted/wt.)	Cost ($ per lb /yd² woven /yd² felt)	Air perm. (cfm/ft² @ 0.5 in. w.g.)
			acid	alkali	oxidizer	organic solvents			
Cotton	180(225)	1.6	P	G	P	G	Sateen	0.40/0.41/ --	10-20
Wool	200(250)	1.3	F	P	P	G	Twill or felt	-- /1.77/3.97	20-60
Nylon	200(250)	1.35	P	G	P	G	2X2 Twill/186X118	1.00/0.70/ --	5-22[b]
Acrylics[c]	240-280	1.12-1.18	G	F	F-G	E	3X1 Twill/76X70	0.80/1.01/4.82	20-45
Nomex	425	1.35-1.38	G-F	E[d]	P	E	Felt/14-16 oz	-- / -- /11.5	25-54
Polyesters[e]	280	1.4	G	F	G	E	3X1 Twill/76X66	1.40/1.04/4.82	10-60
Glass fiber	550	2.54	E	P	G	G	3X1 Twill/55X54	0.60/1.75/ --	10-70
Teflon	450	2.3	E	E	E	E	3X1 Twill	-- /12.0/ --	15-65

[a]P = poor, F = fair, G = good, and E = excellent.

[b]Also 1-1200 in all forms.

[c]Acrylics: Orlon (DuPont); Creslan (American Cyanamid); Acrilan (Chemstrand); Crylor (Rhodia Aceta, France); Dralon T (Farbenfabriken Bayer AG, Germany).

[d]At low temperature.

[e]Polyesters: Dacron (DuPont); Fortrel (Fiber Ind./Celanese); Vycron (Beaunit); Kodel (Eastman Chem. Prod.); Enka (Amer. Enka Corp.).

bags will require. The cost of fabrics makes a long bag life essential to economic application of bag filters. Engineers oftimes
cite a 6-mo bag life as the minimum for satisfactorily employing
the bag filter to a particular job.

The individual threads in the fabric consist of many fibers.
The long filamentous fibers of synthetics are preferred over the
short staple fibers of natural materials. Surface treatment of bag
fabrics is often done to make the bags easier to clean. For example,
fiber glass is treated with silicone or, less frequently, with carbon.
Yarn size is measured in cotton yarn numbers (number of 840-yd
hanks per lb) or for synthetics, the denier (d = 5315/cotton yarn
number = gm/9000 m by definition).

EXAMPLE 20-2: What is the filament diameter for 2d orlon if the
s.g. is 1.15 ?

2d fiber weighs 2 gm/9000 m, by definition.

Weight = density X area X length

$$2 \text{ gm} = 1.15 \text{ gm/cm}^3 \text{ X Area in cm}^2 \text{ X (9000 X 100 cm)}$$

$$\text{Area} = 1.93 \text{ X } 10^{-6} \text{ cm}^2 \qquad \text{Diameter} = 1.57 \text{ X } 10^{-3} \text{ cm}$$

$$= 15.7 \, \mu m$$

Plain weaves (1/1; one over and one under for warp and filler),
twill weaves (2/1 or 3/1), and satin ($\geq$4/1) are variously used for
filters, but the twills are the most popular woven fabrics. For a
given thread count, the permeability or openness of the fabric
increases from plain to twill to satin weaves. Felt fabrics (un-
woven mats of fibers fixed by heating and processing) are often
used for bag filters. The felt may be needled to give additional
bonding of the fibers in the fabric.

b. Fiber Mats or Beds. Fibers may be put into mats or beds
and used as filters without attachment of the fibers to each other.
The fibers, most popularly fiber glass but sometimes slag wool
or other material, may be randomly oriented or they may be combed
or carded to arrange them in parallel.

c. Granular Filters. Beds of granules such as sand are
sometimes used as air filters. Porous ceramics, bonded granules,
may be used in some special application, but they are too expensive for general use. There have been some gravel bed filters;
however, large media such as gravel are used most often in
scrubbers and not alone as filters.

d. Viscous or Wetted Filters. Filters may pass through a
reservoir of oil or have oil sprayed intermittently over their surfaces. The oil washes off the collected dust and provides a
sticky surface for collection. This type of filter is applied princi-
pally to air conditioning. Wetted marble bed filters are described
in Chapter 21.

e. HEPA Filters. HEPA (high efficiency particulate air)
filter material was originally a paper made of 85% esparto fibers
with 15% of Bolivian or South African blue asbestos. The papers
are now made with other filler materials, especially glass and
asbestos fibers.

2. Filter Configurations

Filter media are put into a configuration that will require
the minimum area and volume consistent with operation and maintenance. Space in an industrial plant is at a premium; however,
this factor must not eliminate the accessibility required for ease
of maintenance.

a. Fabric Filters. Fabrics are sewn into tubular shapes
that are sometimes open at the ends, but they are usually closed
at one end with a circular-shaped piece of fabric for the bag or
sewn flat or with an oblong piece for the envelope filter. The
diameter or width of the bag is compatible with the width of the
available fabric, and it normally ranges from 6 to 18 in. in diameter or 12 to 24 in. wide. The bags are usually 6 to 30 ft long,
most popularly 12 to 15 ft for shaker cleaning or shorter for pulse
jet cleaning.

Bags are often fastened to a dirty air plenum (open end down) and suspended from the shaker mechanism at the top of the bag-house, the dirty air passing from the inside of the tube outward (see Fig. 20-3a). Some bag filters and envelope-type filters have the fabric supported by a wire frame and filtration is from the out-side of the bag or envelope inward (see Fig. 20-3b). The out-to-in bags are usually supported open (clean) end up and the enve-lopes are usually horizontal and on edge with the clean air exit at the open end (see Fig. 20-3c). The filters are always arranged such that the dust falls into the hopper. Bag houses with the dirty air plenum at the top and the bags open on both ends have been built, but they have some disadvantages.

Bags are arranged in rows. The bags must be far enough apart (2 in. minimum, usually more) to prevent their rubbing against each other and to permit access for bag replacement; they need to be as close together as practicable to keep the size of the baghouse reasonable. A single section of a baghouse may have only several bags or it may have more than a hundred.

The number of sections in a baghouse is determined largely by operating conditions. As a result, multiple sections should be used to provide filtration, even when one section is off line for cleaning as in some types or for bag replacement.

b. Fiber Mats or Beds. Fiber mat filters are thin mats (1-2 in. thick) of loosely packed (packing density β = few hundredths), randomly oriented fibers. The mat has no structural stability and is held in a container that is perforated or screened to permit air flow. Although the filtration is high (~600 fpm), it is considerably below duct velocities, and the filters may be placed in a sawtooth arrangement to reduce the cross-sectional area space requirement.

Fiber bed filters are thick beds (thickness of feet) of fibers that may be either random or parallel in position. The packing

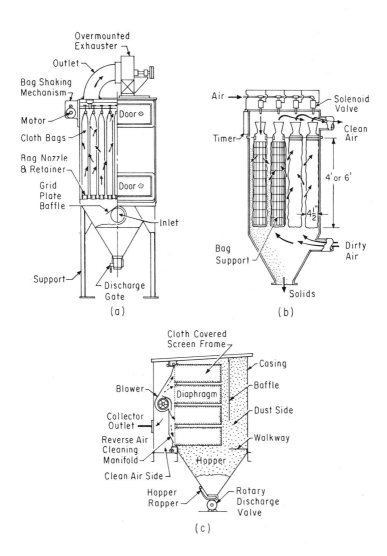

FIG. 20-3. Baghouse configurations. (a) Inside-out filtration/shaker (Pangborn). (b) Outside-in filtration/jet (Mikro-Pulsaire). (c) Envelope/reverse air (Pangborn).

density is usually about 0.1 and the velocities are much lower
than those for the mats.

    c. Granular Filters. Filter beds of sand or other granules
may run up to several feet in thickness and be contained on the
sides by a box shape. The sand is usually supported on layers
of increasingly larger particles with the bottom layer of stones
serving to distribute the air from the ducts over the filter area.
Bonded granular materials with structural integrity can be used
without support except at the edges.

    d. HEPA Filters. HEPA or absolute filters are held in
rigid frames, frequently 2 ft square and 11 1/2 in. thick (1000 cfm).
Other sizes of square and round shapes are available. The paper
matrix (200–250 ft^2) is pleated to give a large area in the small
space; all pleats, front and back, are held apart by corrugated
cardboard, metal, plastic, or other spacers which provide the
channels for the air flow.

    3. Filter Cleaning

    The throwaway filters are not cleaned; they are merely dis-
carded when they get dirty. Throwaway filters include HEPA,
fiber mats, and granular beds. Viscous filters are occasionally
scrubbed to rid them of accumulations. Fabric filters require
frequent cleaning.

    The method of cleaning fabric filters varies according to
manufacturer and type of filter. The filter cake is allowed to
build until the head loss (ΔP) becomes intolerable; then the cake
must be removed. The removal may be accomplished by shaking
or vibrating the bags where they are suspended, but more frequently
nowadays the cloth is moved by increasing the air pressure on the
bag in a manner that will cause the bag to collapse or otherwise
deform enough to dislodge the filter cake. The cleaning air may
be supplied as a jet pulse into the bag, a back pressure of the

exiting clean air, a stream of the clean air that is blown into rows
of bags, or other ways. The frequency of bag cleaning has a direct relation to the amount of material penetrating the filter.

If the head loss after cleaning, the residual head loss,
becomes too large after long use, the bags may be laundered to
remove the adhering particles, then put back into service. Such
adhesion is less likely on treated and un-napped fibers.

B. Design Calculations

Many formulas have been proffered for the calculation of
filtration parameters. A few of them for fabric filters are presented
here for the purpose of illustrating the approaches used to develop
such formulas; they should aid in the understanding of the factors
involved. In the usual bag filter, filtration is accomplished by the
fabric for only a brief period; then the filtration is by the cake of
collected dust. Fortunately, the cake is much more efficient at
filtration than is the fabric.

1. Filter Efficiency

The usual approach to the calculation of filter efficiency
(η) is the projection of probability of collection by a single fiber
(granule) to the probability of collection for the many fibers (granules) in the filter matrix. Dorman applies this technique to obtain
efficiency prediction by

$$-dn/n = 8\beta r v_0 \, db/(\pi Q D_f^2) = 8\beta r \, db/[(1 - \beta)\pi D_f^2], \quad (20-1)$$

where dn/n = change in number of particles of a given size,
β = packing density of filter matrix material = absolute volume of
matrix/volume of filter, r = half-width of collection for gas ap-
proaching cylindrical obstacle = "effective radius" of obstacle,
v_0 = face velocity for filter (also termed superficial velocity)
= Q/A, Q = volumetric flow rate, A = cross-sectional area of
filter normal to flow, D_f = diameter of fiber in filter, and db = incremental thickness of filter in the direction of air flow.

Integration of Eq. 20-1 yields

$$q = n/n_0 = \exp\{-4\beta\eta_s b/[\pi D_f(1 - \beta)]\}, \qquad (20-2)$$

where q = penetration of filter, n/n_0 = number of particles pene-
trating/number of particles entering, and η_s = efficiency of a
single fiber in mat $\neq$ efficiency of single isolated fiber because
the effective length of the fiber is reduced by blinding at fiber
overlaps and the flow regime around the fiber is changed by the
presence of neighboring fibers. The difficulty in obtaining η_s
points up the inexactness of this approach.

EXAMPLE 20-3: What is the efficiency of collection (η) for particles with η_s = 0.05, D_f = 8 μm, β = 0.2, and b = 1/16 in.?

$$q = \exp\{-4(0.2)(0.05)(1/16 \times 25,400)/[3.14(8)(1 - 0.2)]\}$$
$$= \exp(-3.16) = 0.043$$
$$\eta = 1.000 - q = 1.000 - 0.043 = 0.957 \text{ or } 95.7\%$$

Another complicating factor in attempts to make efficiency calculations for fabric filters is that the penetration changes as the
cake builds and the cake filtration is largely a result of interception
by the cake particles and not impaction on the filter fibers.

2. Pressure Drop

The pressure drop (ΔP) is highly important in filtration
economy and, as a result, it has received much attention. Davies
modified an approach introduced by Langmuir to obtain

$$\Delta P = 64\mu Q b\beta^{1.5}(1 + 56\beta^3)/(AD_e{}^2) \quad \text{for } \beta < 0.02, \quad (20-3)$$

and

$$\Delta P = 70\mu Q b\beta^{1.5}(1 + 52\beta^{1.5})/(AD_e{}^2) \quad \text{for large } \beta, \quad (20-4)$$

where ΔP = head loss across filter (dyne/cm^2), μ = gas viscosity
(poise), Q = air flow (cm^3/sec), A = filter face area (cm^2),
D_e = effective fiber diameter (cm; larger than microscopic diameter), b = thickness of filter (cm), and β = packing density.

EXAMPLE 20-4: What constants should replace the 64 and 70 in
   Eqs. 20-3 and 20-4 when ΔP is in. w.g., Q is cfm, A is
   ft^2, and D_f is μm?

   Multiplier: $[4.015 \times 10^{-4}$ in. w.g./(dyne/cm^2)]

   $\times (30.5$ cm/ft)^3/(30.5 cm/ft)$^2 \times (1/60$ min/sec)

   $\times (1/10^{-4}$ cm/μm) = 2.04

   Replacement constants:

   $64 \times 2.04 = 131$ and $70 \times 2.04 = 143$

The increase in ΔP (also called filter drag, S) with cake
buildup is shown in Fig. 20-4. Borgwardt and Durham characterized
the change in ΔP in three regimes of flow by

$$d(\Delta P)/dV = K (\Delta P)^n, \qquad (20-5)$$

where V = volume of gas filtered, n = parameter of fit = 2 for
blocking and straining = 1.5 for depth or intermediate filtration
= 0 for cake filtration, and K = constant. It will be noted that

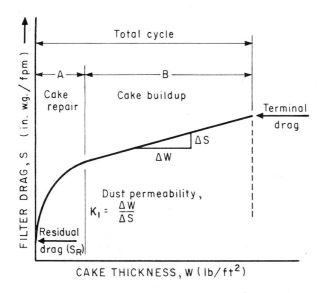

FIG. 20-4. Head loss or drag versus cake thickness.

cake filtration simply follows Darcy's law for flow through porous media; i.e.,

$$v = K \Delta P/(\mu b) ,\qquad\qquad (20\text{-}6)$$

where v = superficial velocity. Dust permeability (K_1), as defined in Fig. 20-4, is used to determine the frequency of cleaning required.

Total ΔP includes not only the head loss across the filter but also the head losses incurred in conducting the gas stream to and away from the filter.

Obviously, when just cleaned filters are in parallel with other filters having varying cake thicknesses, the head loss across all the filters will be the same but the flow rates through the different filters will vary. Figure 20-5a shows the air velocities resulting from operating six component elements (compartments, rows, or individual bags) in parallel with one component cleaned at a time. Flow rate will follow a curve such as that in Fig. 20-5b for a multicompartment, intermittently cleaned filter house. In the usual case the discharge of the fan is an inverse function of the pressure loss, and both flow rate and head loss vary with time. Large baghouses and modern baghouses with continuous cleaning show practically constant heads and flow rates.

3. Quality of Filter

The relation between penetration and head loss has been used as a basis for defining a quantitative unit for comparison of filters:

$$Q_f = - (100 \log_{10} q)/\Delta P ,\qquad\qquad (20\text{-}7)$$

where Q_f = quality factor of filter, q = penetrating fraction, and ΔP = head loss (in. w.g.).

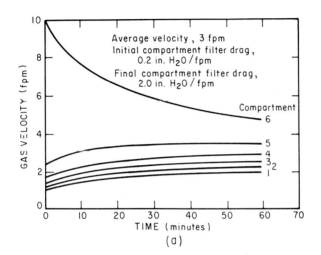

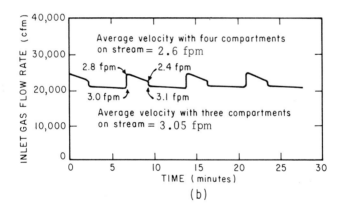

FIG. 20-5. Operating parameters for multicompartment,
intermittently cleaned baghouse. (a) Velocities after cleaning
compartment 6. (b) Gas flow rates for cleaning one-third of
compartments at a time. From G. W. Walsh and P. W. Spaite,
"An Analysis of Mechanical Shaking Air Filtration, " J. Air Pollution
Control Assoc., 12:2, 57-61, February 1962.

C. Selection Criteria

Filters can be used to remove any size particles with any specified efficiency; therefore, the cost of cleaning is the ultimate basis of whether to use filtration. The minimum annual cost of a filter installation is obtained by tradeoffs among the filtration velocity, the penetrating fraction of the dust in question, and the pressure loss. All three parameters increase simultaneously.

The operating temperature of the bag filter is very critical; the temperature must be kept above the dew point to prevent wetting and blinding of the filter and acid formation and below the temperature that will degrade the fabric. These two limits were often rather close together (less than $100^\circ F$ apart) before the advent of high temperature filter materials, first fiberglass and then high temperature nylon (HT-1 or Nomex).

Fabric filter velocities (air-to-cloth ratios) range from 1/2 to 30 fpm with most applications in the 2 to 6 fpm bracket. The lowest velocities are used for uniform, small particles such as lead fumes and the highest for coarse, nonabrasive particles such as wood dust. In view of present emission limits and continuous bag cleaning with reverse jets of air, the most satisfactory velocities for typical applications appear to be 4 or 5 fpm. Filter velocities have increased over the years with improvements in bag filters (see Fig. 20-6).

Head loss on new filter fabric is quite low; the drag (S) is normally on the order of a few to several hundredths in. w.g./fpm. The residual drag after a bag has undergone many cleanings is usually within a factor 2 of 0.5 in. w.g. If the residual drag becomes excessive, more than about an inch, the bag may be washed.

IV. APPLICATIONS

Filters are used to separate dusts and sometimes mists from aerosol streams in many types of industries. In some rare in-

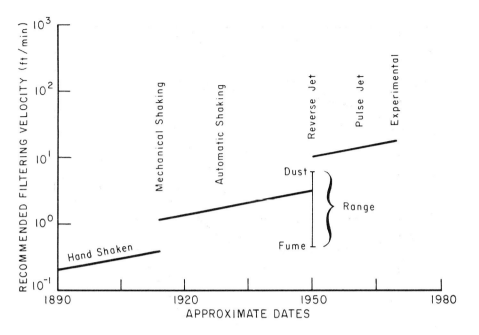

FIG. 20-6. Filter velocities increase. Adapted from C. E.
Billings and J. Wilder, Handbook of Fabric Filter Technology,
Vols. I-IV, GCA Corp. Report to National Air Pollution Control
Administration, Bedford, Massachusetts, December 1970.

stances, chemically impregnated filters may be used to absorb a
gaseous component of a waste stream. The use of a filter in a
given application should be decided by the characteristics of the
gas streams and the particles and the applicability of a filter to
such conditions.

Industrial filtration is usually done with fabric or bag filters.
Bag filters have the following advantages: (i) can attain low penetration, high efficiency, even on rather small particles (see Fig.
20-7); (ii) can collect material in the dry state for easy return to
the process or for avoiding corrosion and/or water pollution problems; (iii) may be operated in the presence of flammable or explosive materials; and (iv) are manufactured and marketed by a

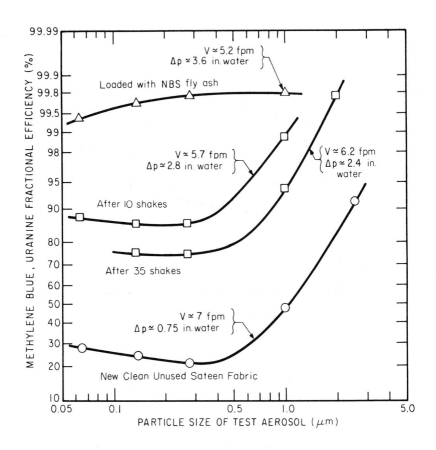

FIG. 20-7. Filter penetration with and without cake. From K. T. Whitby and D. A. Lundgren, <u>Fractional Efficiency Characteristics of a Torit Unit-Type Cloth Collector</u>, Torit Mfg. Tech. Rept., Minneapolis, Minnesota, August 1961.

highly competitive industry. Fabric filter drawbacks include: (i) cannot tolerate abrasive materials that shorten bag life; (ii) may be ruined by high temperatures; (iii) may be attacked by acid or alkali materials; (iv) may reach explosive concentrations of dust in the housing; (v) may be blinded and ruined by sticky materials; and (vi) may require heating or flushing of baghouse to prevent acid condensation when shutting down.

Some of the successful applications for bag filters include cement manufacture (product and kiln dust), lime manufacture (product and kiln dust), gypsum, limestone, soaps and detergents, wood products, grain dusts, alfalfa dust, ceramics, chocolate, pesticides, talc, tobacco, carbon black, baking powder, fertilizer, steelmaking, lead smelting, and hot-mix asphalt plants. Most unsuccessful attempts to use baghouses have been with sticky or abrasive particles.

In 1969 the relative popularities of various filter fabrics were as follows: cotton, 33%; glass, 33%; polyester (Dacron, etc.), 15%; orlon, 5%; wool, 5%; and other, 9%. High temperature nylon (Nomex) is rapidly increasing in popularity, despite its rather high cost, because it has the highly desirable properties of high temperature operation and long bag life.

Deep bed filters have been used for the diffusional collection of small particles. A bed several feet thick was designed for use at the Hanford, Washington, AEC facility to catch radioactive particles. Mist eliminators have been used to remove acid mists and oil mists both in relatively thick beds (up to several inches) and in thin impaction beds.

V. COSTS

The installed equipment costs for fabric filters are usually intermediate among the costs for high efficiency removal of fine particulates, higher than venturi scrubbers and lower than electrostatic precipitators for the same applications; the usual costs per cfm capacity for the three units being about $2, $3.5, and $6, respectively. However, the total annual costs for fabric filters, in suitable applications, are much less than those for other high efficiency cleaners (see Fig. IV, p. 154).

The fabric filters with continuous automatic cleaning cost more than the filters with intermittent shaking; much of this cost difference can be justified on the basis of the higher filtration velocities possible with the automatic cleaning. For example, the superficial velocities on comparable dusts are usually about 5 fpm and 2 fpm for the two types of cleaners above; costs of filters vary approximately with size to the 0.9 power (rather than the usual 0.6 cited in Chapter 12); $(5/2)^{0.9} = 2.28$ or the automatically cleaned filter can cost 2.28 times as much per square foot of cloth and still not be any more expensive than the intermittent shaker type.

The big item in filter-operating costs is the bag life; the lifetime sets the fabric and installation costs. The Billings and Wilder reference shows the relative costs for different filter materials as follows: cotton, 1 (base); wool, 2.75; orlon, 2.75; Dacron, 2.8; fiberglass, 5.5; and Nomex, 8. The typical bag material is $2/yd^2 and costs about $4.5 for a 20-ft^2 bag; the bag itself costs about $10.

VI. SUMMARY

Particles of any size above the size of molecules can be removed from an aerosol stream by filtration. The usual filter matrices for various filter applications are fabric for industrial dusts, fiber mats or beds for mist eliminators and air conditioning, and asbestos paper for ultra-high efficiency (HEPA).

The bag filter is the most economical high efficiency collector for those applications to which it is suited. Bag filters should not be used for collecting sticky or highly abrasive particles. Cooling of the gas stream is often required before fabric filtration.

PROBLEMS

1. Why would a bag filter not be suitable for each of the following applications: smoke from a meat smokehouse, particles from lime slaking, fly ash from coal burning, radioactive fumes, and heavy mist from a chemical process?

2. What is the total weight of a filter bag 30 ft long X 18 in. in diameter with a 1/16-in. cake of lead fumes (ρ = 9.3 gm/cm^3) with 0.4 voids fraction if the fabric is 18 oz/yd^2?

3. What is the estimated cost of replacing 2000 bags of 24 ft^2 each with 14-oz Nomex bags if the labor is 20 min/bag?

4. Excluding the collection and storage hoppers, compare the volume of a baghouse that is 18 ft high, has 20 ft^2 of fabric per ft^2 of plenum (plan area), and uses an air-to-cloth ratio of 5 ft/min with the volume of an electrostatic precipitator to achieve 99.5% collection with an effective migration velocity (ω) of 0.33 ft/sec and a migration distance (s) of 4.8 in. if the units are to clean 400,000 acfm. For electrostatic precipitators, $q = \exp(-\omega t/s)$ from Chapter 22.

5. What are the approximate dimensions for the cleaners in Problem 4 if the width is about twice the height and the length is sufficient to give the volume calculated? What are the retention times (t) in seconds for the two units?

6. What is the optimum air-to-cloth ratio for a bag filter if at 50,000 acfm the installed cost = size$^{0.9}$ (K) = \$2.5 for air-to-cloth of 5 fpm which gives average head loss of 6 in. w.g. and the head varies directly with the filtration velocity? Assume electricity is \$0.015 per kWh and that bag replacement and maintenance costs are constant.

BIBLIOGRAPHY

Air Pollution Manual, Part II: Control Equipment, Amer. Industrial Hyg. Assoc., Detroit, Michigan, 1968.

Billings, C. E., and J. Wilder, Handbook of Fabric Filter Technology, Vols. I-IV, GCA Corp. Report to National Air Pollution Control Administration, Bedford, Massachusetts, December 1970.

Borgwardt, R. H., and J. F. Durham, "Factors Affecting the Performance of Fabric Filters," Paper 29c at 60th AIChE Meeting, New York, November 1967.

Brief, R. S., A. H. Rose, Jr., and D. G. Stephan, "Properties and Control of Electric-Arc Steel Furnace Fumes," J. Air Pollution Control Assoc., 6:4, February 1957.

188   AIR POLLUTION

Davies, C. N., "The Separation of Airborne Dust and Particles,"
Proc. Inst. Mech. Engr. (London), 1B:5, 185-213, 1952.

Dorman, R. G., "Filtration," Aerosol Science (C. N. Davies, ed.),
Chap. VIII, Academic Press, New York, 1966.

Dorman, R. G., "Theory of Fibrous Filtration," High-Efficiency
Air Filtration, Butterworth, London, 1964.

Iinoya, K., and C. Orr, Jr., "Source Control by Filtration," Air
Pollution, Vol. III (A. C. Stern, ed.), Chap. 44, Academic
Press, New York, 1968.

Löffler, F., "Collection of Particles by Fiber Filters," Air Pollution
Control, Part I (W. Strauss, ed.), Wiley-Interscience,
New York, 1971.

"Nomex Filtration," DuPont Bulletin, Geneva, Switzerland, 1971.

Paretsky, L., L. Theodore, R. Pfeffer, and A. M. Squires, "Panel
Bed Filters for Simultaneous Removal of Fly Ash and Sulfur
Dioxide: II. Filtration of Dilute Aerosols by Sand Beds,"
J. Air Pollution Control Assoc., 21:4, 204-209, April 1971.

Phillips, P. E., "Design, Operation and Maintenance of the Baghouse Installation on the Electric Furnaces at Kansas City
and Houston," Paper 64-64, Air Pollution Control Association Meeting, Houston, 1964.

Robinson, J. W., R. E. Harrington, and P. W. Spaite, "A New
Method for Analysis of Multicompartmented Fabric Filtration,"
Atmospheric Environ., 1, 499-508, 1967.

"Selecting Fabrics for Filtration and Dust Collection," J. P.
Stevens & Co., Inc., New York.

Spaite, P. W., and G. W. Walsh, "The Effect of Fabric Structure
on Filter Performance," Presented at American Industrial
Hygiene Conference, Washington, D. C., May 1962.

Stephan, D. G., G. W. Walsh, and R. A. Herrick, "Concepts in
Fabric Air Filtration," Amer. Industrial Hyg. Assoc. J., 21:1,
February 1960.

Strauss, W., Industrial Gas Cleaning, Pergamon Press, New York,
1966.

Walsh, G. W., and P. W. Spaite, "An Analysis of Mechanical
Shaking in Air Filtration," J. Air Pollution Control Assoc.,
12:2, 57-61, February 1962.

White, P. A. F., and S. E. Smith, High-Efficiency Air Filtration,
Butterworth, London, 1964.

Chapter 21

SCRUBBING

I. INTRODUCTION

The liquid scrubbing of particles has always occupied an important position in the removal of particles both in the ambient air and in the gas-cleaning processes of industry. The popularity of scrubbers is probably a result of their innovative design possibilities, relatively low first costs, and small particle collection capabilities.

II. THEORY

Scrubbers increase the particle size with a scrubbing liquid, nearly always water, in order to make collection of the particles easier. The entrainment of the particles (solid and liquid) in droplets may be by either or both of two means, namely, impaction of the water droplet and the particle or condensation of water vapor on the particle. Liquid films may be used to remove particles collected by another means such as electrostatic precipitation, but the removal mechanism is not the scrubbing action considered here.

Droplets falling under the force of gravity are most efficient at particle entrainment when the droplet size is about 800 µm (see Fig. 21-1). However, for equal relative velocities between droplet and particle, small droplet sizes are more efficient than large, as is indicated in the impaction parameter ψ. For example, 100-µm droplets are preferred when the droplets and particles are in a centrifugal field of 100 g. The droplets must be large enough to prevent their evaporation to a size that will carry through the collector and in cocurrent scrubbing large enough to require significant time of acceleration or deceleration of droplets, the time

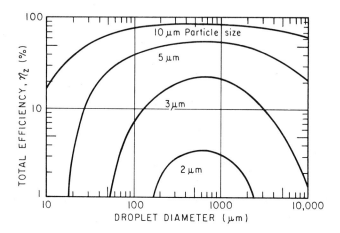

FIG. 21-1. Target efficiency for water droplets under gravity. From C. J. Stairmand, "The Design and Performance of Modern Gas-Cleaning Equipment," <u>J. Inst. Fuels</u> (London), <u>29</u>, 58, February 1956.

during which impaction occurs in the cocurrent scrubber.

Condensation is believed to contribute little to the overall efficiency of the usual scrubber. The current applications of hot water and steam in ejector scrubbers may indicate an increasing importance of condensation in particle scrubbing for fine particles.

III. DESIGN

A simplistic approach to scrubber design calculations can be based on target theory. Each droplet cleans a path with a cross-sectional area that is in the ratio of the individual droplet collection efficiency (η_I) to the projected area of the droplet; i.e., a droplet has an effective area.

Semrau has shown that the overall efficiency of collection by the various types of scrubbers for a given particulate can be related to the contacting power between the liquid and the gas by

$$N_t = \alpha P^\beta ,$$

(21-1)

where N_t = number of transfer units $(q = \exp -N_t)$, P = contacting power (HP/1000 cfm) = $0.157 \Delta h + 0.583 p_L L'$, Δh = head loss across wet part of scrubber (in. w.g.), p_L = pumping pressure of liquid (psig), L' = liquid flow rate (gal/1000 ft^3 of gas), and α and β are parameters of fit. Values of α and β depend on the particle characteristics, especially particle size and size distribution and to some extent on wettability. Some typical values of α and β are listed in Table 21-1.

A. Spray Scrubbers

The target efficiency approach for spray scrubbers gives

$$\eta_t = 1 - q = V_{ac}/V_a = nA_e s/(Q_a t) , \qquad (21-2)$$

where η_t = theoretical fractional efficiency of collection for a given ψ, q = penetration of particles = concentration of particles in outlet of scrubber divided by concentration in the inlet, V_{ac}/V_a = volume of air cleaned/total volume, n = number of droplets in cleaner at any time, A_e = effective cross-sectional area of a droplet, s = distance droplet moves relative to the air, Q_a = gas flow

TABLE 21-1

Coefficients for Correlating Contacting Power with Efficiency[a]

Application	Coefficients	
	α	β
Venturi on lime kiln dust	1.47	1.05
Venturi on black liquor recovery furnace	1.75	0.620
Venturi on phosphoric acid mist	1.33	0.647
Spray cyclone on talc dust	1.16	0.655
Venturi on odorous mist	0.363	1.41
Venturi on open-hearth steel furnace	1.26	0.569

[a]From K. T. Semrau, "Correlation of Dust Scrubber Efficiency," J. Air Pollution Control Assoc., 10:3, 200-207, March 1960.

rate, and t = holdup time or time that a droplet remains in scrubber. Furthermore,

$$n = Q_w t/(\pi D^3/6); \quad A_e = \eta_I A; \quad \text{and } s = v_g t , \qquad (21\text{-}3)$$

where Q_w = water flow rate, t = holdup time = Z/v_d, Z = length of droplet path, v_d = droplet velocity relative to cleaner in Z direction, D = diameter of droplet, η_I = target impaction efficiency (see Fig. 20-2; $q_I \approx \exp -\psi$ for $\psi < 1 \approx \exp -\sqrt{\psi}$ for $\psi > 1$), A = actual projected droplet area = $\pi D^2/4$, and v_g = gas stream velocity relative to scrubber. Substitution of Eqs. 21-3 into Eq. 21-2 gives

$$\eta_t = [3 v_g t \eta_I/(2 D)] (Q_w/Q_a) . \qquad (21\text{-}4)$$

Because the random paths cleaned by the droplets overlap (the value of η_t may exceed unity), the expression is not the actual efficiency but represents the number of transfer units (N_t) as used above and described in Chapter 12. The calculated efficiency is

$$\eta_0 = 1 - q = 1 - \exp(-N_t) = 1 - \exp(-\eta_t) \qquad (21\text{-}5)$$

or with the units adjusted so that v_g is ft/sec, t is sec, D is μm, and Q_w/Q_a is gal/1000 acf of gas,

$$q = \exp (-61.2 v_g tL'\eta_I/D) . \qquad (21\text{-}6)$$

For countercurrent spray towers, Z = height of tower (H) and the relative velocity for calculating ψ is the terminal settling velocity for the droplets minus the upward velocity of the gas. For crossflow towers, the slant path of the particles and the velocity of the droplet are both increased by the secant of the slant angle; however, the relative velocity is simply the terminal settling velocity of the droplets (neglecting acceleration) and the holdup time is merely the height over the terminal settling velocity.

EXAMPLE 21-1: What is the calculated penetration for a spray tower with a water droplet diameter of 500 μm and particles 5 um in diameter with a density of 2.0 gm/cm^3 if the gas velocity in the tower is 3 fps (upward) and the tower height is 30 ft? Use a liquid flow rate of 1.5 gal/1000 ft^3 and a temperature of 20°C.

Droplet velocity: v_d = terminal settling velocity

minus gas velocity = $u_t - v_g$

$u_t = 0.34 \, \rho_p^{2/3} d_u$ Intermediate regime; Eq. 5-44

$\quad = 0.34 (1.0)^{2/3} (500) = 170$ cm/sec = 5.58 ft/sec

$v_d = 5.58 - 3.00 = 2.58$ ft/sec = 78.6 cm/sec

Holdup time: 30 ft/2.58 ft/sec = 11.6 sec

Target efficiency:

$\psi = C_c \rho_p d_p^2 v_r / (18 \mu D)$ $C_c = 1 + 1.7 \lambda / d$
$\qquad\qquad\qquad\qquad\qquad\quad = 1 + 1.7 (0.063)/5 = 1.02$

$\quad = 1.02 (2.0) (5 \times 10^{-4})^2 (78.6) / [1.8 \times 10^{-4} (18) (500 \times 10^{-4})]$

$\quad = 0.248$

$q \approx \exp(-0.248) = 0.78$ $\eta_I = 0.22$ (see also Fig. 20-2)

Tower penetration:

$q = \exp[-61.2 (0.22) (3) (11.6) (1.5)/500] = 0.25$ or 25%

Spray nozzles are normally under 20-100 psig (fog sprays, 300-450 psig), have liquid flow rates of 0.5-2 gal/1000 ft^3 of gas, and use sprays introduced in varying orientations to the gas flow-- normal, parallel, or tangential.

An approximation for a cyclonic scrubber with tangential, bottom entry of the gas stream replaces the t in Eq. 21-4 with the inlet width b divided by the radial velocity of the droplet. Cyclone scrubbers are available in capacities of 500->50,000 cfm with gas velocities to 200 fps, separation factors 50-300 g, liquid flow rates 3-10 gal/1000 ft^3 of gas, and head losses 2-6 in. w.g.

B. Venturi Scrubbers

Venturi scrubber analysis and design depend on the droplet size created by shear forces of the gas stream on the water. Nukiyama and Tanasawa reported the droplet size from gas atomization as

$$D = (585/v_r)\sqrt{\sigma/\rho_L} + 597 (\mu_L/\sqrt{\sigma \rho_L})^{0.45} (1000 \, Q_L/Q_g)^{1.5}, \quad (21\text{-}7)$$

where D = droplet diameter (μm), σ = surface tension of liquid (dynes/cm), ρ_L = density of liquid (gm/cm^3), μ_L = dynamic viscosity of liquid (poise), v_r = relative velocity of gas to liquid (m/sec), and Q_L/Q_g = ratio of liquid flow rate to gas flow rate. For water and air at NTP, Eq. 21-7 becomes

$$D = 16,400/v_r + 1.4\,L'^{1.5}, \qquad (21-8)$$

where L' = liquid flow rate (gal/1000 ft^3 of air) and v_r is ft/sec.

Johnstone and Roberts correlated dust collection in a venturi scrubber (throat velocities to 240 ft/sec) with the specific surface of the droplets according to

$$q = \exp(-KS), \qquad (21-9)$$

where q = penetration, S = specific surface of droplets (ft^2/ft^3 of gas) = 245 L'/D, and $K \simeq 0.125$ for dust (see Fig. 21-2). Johnstone et al. reasoned that since impaction is the major collection mechanism, collection should relate to the impaction parameter (ψ). They found

$$q = \exp(-KL'\sqrt{\psi}), \qquad (21-10)$$

where $K \simeq 0.09$ for particle collection.

For usual conditions, the collection efficiency (for a given head loss) on a probability scale has been depicted as a straight line versus particle size on a log scale with an indicated σ_g of 5.5. The 50% cut size (d_{50}) at various head losses may be calculated from

$$d_{50} = \exp[-1.56 - 1.46\,\ln(\Delta h/5)], \qquad (21-11)$$

where d_{50} = size of particle collected with 50% efficiency (μm) and Δh = head loss in the venturi scrubber (in. w.g.).

EXAMPLE 21-2: What is the d_{50} for 20 in. w.g. head loss in a venturi scrubber? What size will be collected with 97.5% efficiency?

$$d_{50} = \exp[-1.56 - 1.46\,\ln(20/5)] = 0.028\,\mu m$$

$d_{97.5}$: 2.5% excluded (one-tail) in $\overline{x_g} \times \sigma_g{}^K$ when K = 1.96, i.e., at 95% confidence limit.

$$d_{97.5} = d_{50} \times \sigma_g{}^{1.96} = 0.028(28.3) = 0.78\,\mu m$$

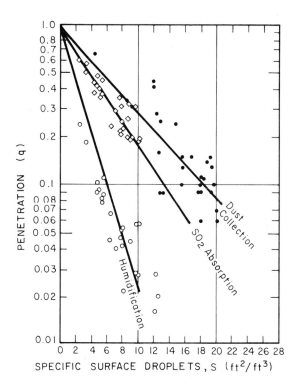

SPECIFIC SURFACE DROPLETS, S (ft^2/ft^3)

FIG. 21-2. Penetration versus specific surface of droplets. From H. F. Johnstone and M. H. Roberts, "Deposition of Aerosol Particles from Moving Gas Streams," Ind. Engr. Chem., 41:12, 2417-2423, December 1949.

Note: This approach is probably good; however, the calculated sizes appear somewhat small. P in Eq. 21-1 is
equal $\Delta h/3.81$ for 60% fan efficiency.

Calvert et al. have derived an equation for venturi scrubbing
of particles based on

$$dc/c = (2/55) D\rho_L v_g (Q_w/Q_a) df, \qquad (21-12)$$

where c = concentration of particles, f = ratio of v_r/v_g, and other
terms are as used above. The equation they arrived at is

$$q = \exp\left\{\frac{2 D\rho_L v_g}{55\mu} \frac{Q_w}{Q_a} \frac{1}{\psi}\left[-0.7 - \psi f + 1.4\,\ell n\left(\frac{\psi f + 0.7}{0.7}\right) + \frac{0.49}{\psi f + 0.7}\right]\right\},$$

$$(21-13)$$

where ψ = impaction parameter and $f \simeq 0.45$ for hydrophilic particles with $L' > 5 \simeq 0.2$ for hydrophobic particles with $L' > 2$. Each of the fractions in Eq. 21-13 should be dimensionless.

EXAMPLE 21-3: What is the q for 0.3 μm-diameter particles with $\rho_p = 2.5$ gm/cm^3, $v_r = 300$ fps (91.5 m/sec), $L' = 5.5$ gal per 1000 ft^3 as calculated by Eqs. 21-9, 21-10, and 21-13?

$$D = 16,400/300 + 1.4(5.5)^{1.5} = 54.67 + 18.06 = 72.7 \text{ μm}$$

q by Eq. 21-9:

$$q = \exp(-0.125 \times 245 \times 5.5/72.7) = \exp(-2.32) = 0.10$$

q by Eq. 21-10:

$$\psi = \frac{1.36(2.5)(9150)(0.3 \times 10^{-4})^2}{18(1.8 \times 10^{-4})(72.7 \times 10^{-4})} = 1.19$$

$$q = \exp(-0.09 \times 5.5\sqrt{1.19}) = 0.58$$

q by Eq. 21-13:

$$q = \exp\left\{ \frac{2(72.7 \times 10^{-4})(1)(9150)}{55(1.8 \times 10^{-4})} \frac{5.5}{7480} \frac{1}{1.19} \left[-0.7 - 1.19(0.45) \right.\right.$$
$$\left.\left. + 1.4 \ln\left(\frac{1.19(0.45) + 0.7}{0.7}\right) + \frac{0.49}{0.7 + 1.19(0.45)} \right]\right\}$$

$$= \exp(8.30 \times -0.0435) = 0.70$$

Note: Equation 21-9 was developed on larger, mixed dust sizes. It is obvious that q is not independent of d_p. Equation 21-13 is very unstable because of term structure and size of terms.

Gieseke used classic transfer principles and the inertial impaction concept to obtain the following relation:

$$N_t = (C\eta_I N_{Re}^{0.5} \Delta h)/(\rho_g v_g C_p^{2}), \tag{21-14}$$

where N_t = number of transfer units, C = constant, η_I = target efficiency, N_{Re} = Reynolds number, Δh = head loss, ρ_g = density of gas, v_g = gas velocity relative to cleaner, and C_p = heat capacity of liquid. This approach is equivalent to having all of the β values for Eq. 21-1 equal to one.

The head loss in a venturi scrubber may be calculated as two parts, the fraction of the throat velocity head that is lost in accelerating and decelerating the gas plus the head that is required to accelerate the liquid. The loss is approximated by

$$\Delta h = 4 \times 10^{-5} v_g^2 L' , \qquad (21\text{-}15)$$

where Δh = head loss (in. w.g.). Since L' does not vary widely
in usual design practice, a first approximation of the head loss
is simply a constant times v_g^2. For L' = 5.6 gal/1000 ft^3 of air,
a reasonable value, the head loss by Eq. 21-15 becomes the
velocity head for the throat velocity; i.e.,

$$\Delta h = (v_g/66.7)^2 . \qquad (21\text{-}16)$$

A logical inference from Eq. 21-16 is that the head recovered in
the outlet transition of the venturi is just enough to accelerate
the water added.

Venturi scrubbers have different throat shapes (see Fig. 21-3).
The most popular types now have adjustable throats so that the
velocity can be maintained constant during fluctuating flow rates.
The adjustments may be actually changing the cross-sectional
area of the throat by squeezing in the sides of a rubber throat,
but the adjustment is usually obtained by a variable annulus with
a disk or cone in a tapered section or a rotatable wedge in a rectangular section. The adjustable throat also permits increasing
the throat velocity (head loss) and thereby the efficiency for a
given flow.

Venturi scrubbers range in size from 200-300,000 cfm, in
throat velocities from 150->600 fps, in head losses from 10->80
in. w.g., and in typical liquid flow rates from 5-7 gal/1000 ft^3
of gas with liquid pressures of 5-20 psig.

C. Ejector Venturi

The ejector venturi differs from the conventional venturi
scrubber in that the scrubbing fluid is injected with sufficient
power to pump the gas (see Fig. 21-3); therefore, there is a gain
in pressure (draft available) across the scrubber instead of a

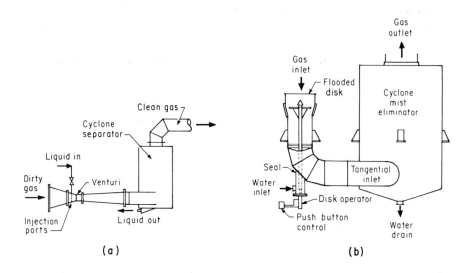

(a)   (b)

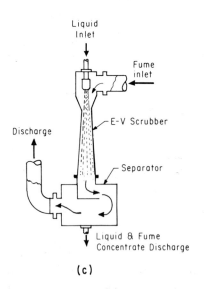

(c)

FIG. 21-3. High-energy scrubbers. (a) Conventional venturi, (b) flooded disk scrubber, and (c) ejector venturi.

pressure drop. The ejector venturi pumps large volumes of water, nearly 100 gal/1000 ft^3 of gas for 5 in. w.g. draft. The penetration is approximated by

$$q = \exp(-30.5\,\eta_I \Lambda L'/D) , \qquad\qquad (21-17)$$

where Λ = effective length of scrubber (ft) and other terms as used above.

Current developments in this field include cleaning systems which use hot water and steam for the powering fluid. The ADTEC system uses hot exhaust gases (>800°F) to heat water (0.8-1.0 lb of water/lb of gas) to 350°-400°F and a pressure of 275-300 psig. The hot water passes through a nozzle, where about 15% flashes to steam, and contacts the 100-fps gas stream with a velocity of about 1000 fps. The gas-water mix (150°F) is separated in a cyclone; the clean air is discharged and the dirty water treated for recirculation. The water treatment for recirculation to the heat exchanger appears to be the major stumbling block in the system, especially the sulfates removal. The scrubber reportedly achieves 99.6%$_w$ efficiency on a dust with 85%$_w$ less than 0.1 μm in diameter.

D. Other Scrubber Types

The impingement baffle plate scrubber (Peabody) humidifies the gas stream with sprays, then passes the stream through a flooded, perforated plate for scrubbing. The plate has 600-3000 holes/ft^2 with impingement plates located above the holes. The gas velocity through the holes, about 100 fps, shears droplets from the adjacent liquid; these droplets do the scrubbing.

The induced spray scrubbers pass the dirty gas stream through slightly submerged slots. The passage of the air violently agitates the scrubbing liquid, forms droplets, and scrubs the particles. The variations in design simply vary in the configuration of the slots.

Marble bed scrubbers are, as the name implies, wetted
layers of marbles up to several inches in thickness. The velocity
through the bed is kept high enough to keep the bed moving in
order to prevent scale buildup on the marbles. The bed material
may be marbles of fiberglass or other material or may be other
round smooth objects, including ping pong balls.

IV. APPLICATIONS

Low-energy scrubbers are capable of achieving the low
penetration limits prescribed by the strict emission regulations
of today only on rather coarse dusts, for example, the kiln dust
from the drier of a hot-mix asphalt plant operating with washed
aggregate. High-energy scrubbers can be made to remove almost
any size particles; however, the removal of very small particles
is effected only with a large expenditure of contacting power.

The suitability of scrubbers for various applications may be
indicated by the relative advantages and disadvantages of scrubbers
when compared with other particulate removing processes. The
principal advantages of scrubbers are low first cost, ability to
handle sticky particles, safety in handling explosive or flammable
particles or gas streams, flexibility in design for fluctuating flows
and/or changing efficiencies, and possibility of simultaneous
absorption of a gaseous pollutant. The major disadvantages of
scrubbers are high operating costs, wet collection of particles
with attendant problems of material reuse or disposal and corrosion, solubility, and scaling, and noise of high velocity gases.

Scrubbers have been used to control dusts from crushing,
drying, and handling operations in lime plants, ore milling, fertilizer plants, quarries, hot-mix asphalt plants, cement batch
plants, and detergent manufacturing; fumes from smelters, cupolas,
and foundries; mists from acid making, electroplating, and pickling;
fly ash from coal burning; smoke from incineration; and odors from
precipitation, sticky particles usually require irrigation of the

rendering operations, kraft paper making, and roofing manufacturing.

Packed towers will collect particles, but plugging difficulties usually preclude the use of such towers for any particles except mists or very soluble particles.

One factor that needs careful attention in the application of scrubbers is that the unit will collect the particle size required by the application. The variable throat venturi can collect small particles if sufficient head is available; therefore, it is usually good practice to buy ample fan capacity for increasing the efficiency. This is especially desirable since the actual power requirements are nearly always underestimated in the preliminary selection of high-energy scrubbers.

V. COSTS

The purchase cost for a scrubber with about 20,000 acfm capacity is likely to be in the range of $0.30 to $1 per cfm. For low-energy scrubbers the annual cleaning costs for the amortized lifetime of the equipment will be on the order of $0.30 to $0.50 per cfm. For high-energy scrubbing the cost will probably be two or three times as high as for the low-energy installation. It will be noted that the low first cost is more than offset in many cases by the high operating costs (see Table IV, p. 152, and Fig. IV, p. 154); therefore, economic application of scrubbing may result from the unique suitability of the scrubber to the given application rather than a competitive edge in costs.

VI. SUMMARY

Scrubbers find wide application in air cleaning. They are capable of cleaning all particle sizes to the degree desired. Scrubbers are the cleaners of choice for sticky particles; such particles cannot be removed by filtration, and for electrostatic electrodes, i.e., wet collection.

Most of the scrubbing action in air cleaners is inertial
impaction between the droplets and the particles. Theoretically,
the condensation of the scrubbing liquid on the particles is attractive, but the contribution of this mechanism to the overall
scrubbing process is small. As a result, the efficiencies attained
by scrubbers bear a simple functional relationship to the contacting
power in the scrubber.

PROBLEMS

1. Plot the target efficiency (η_I) for 700-μm water droplets versus
   particle size (d_p) by taking the values from Fig. 21-1. Fit
   a formula to the η_I versus d_p relationship.

2. Make a log-log plot of Eq. 21-1 for the dusts listed in Table
   21-1.

3. Briefly describe the logical relationships between particle size
   and size dispersion and the position and slope of a line in
   the Problem 2 plot.

4. How well does Eq. 21-6 agree with Fig. 21-1?

5. What effect does a temperature increase from 20° to 80°C have
   on the calculated water droplet size at 400 fps relative velocity and 6 gal/1000 ft^3 liquid flow rate?
   At 20°C, $\sigma = 72.8$ dynes/cm and $\mu_L = 1.005$ cp, and at 80°C,
   $\sigma = 62.6$ dynes/cm and $\mu_L = 0.357$ cp.

6. Calculate the penetration of 1 μm-diameter particles with a
   density of 2.0 gm/cm^3 for a venturi scrubber with a throat
   velocity of 400 fps, a liquid flow rate of 6 gal/1000 ft^3 of
   gas, and a temperature of 80°C by the various formulas,
   including contacting power.

7. What conclusions should be drawn from comparing the calculated penetrations in Problem 6?

BIBLIOGRAPHY

Air Pollution Manual, Part II: Control Equipment, Amer. Industrial
   Hyg. Assoc., Detroit, Michigan, 1968.

Calvert, S., "Source Control by Liquid Scrubbing," Air Pollution,
   Vol. III (A. C. Stern, ed.), Chap. 46, Academic Press,
   New York, 1968.

Calvert, S., D. Lundgren, and D. S. Mehta, "Venturi Scrubber Performance," J. Air Pollution Control Assoc., 22:7, 529-532, July 1972.

Harris, L. S., "Energy and Efficiency Characteristics of the Ejector Venturi Scrubber," J. Air Pollution Control Assoc., 15:7, 302-305, July 1965.

Johnstone, H. F., R. S. Field, and M. C. Tassler, "Gas Absorption and Aerosol Collection in a Venturi Atomizer," Ind. Engr. Chem., 46:8, 1601-1608, August 1954.

Johnstone, H. F., and M. H. Roberts, "Deposition of Aerosol Particles from Moving Gas Streams," Ind. Engr. Chem., 41:12, 2417-2423, December 1949.

Kopita, R., "The Use of an Impingement Baffle Scrubber in Gas Cleaning and Absorption," Air Repair, 4:4, 219-223, February 1955.

Lancaster, B. W., and W. Strauss, "Condensation Effects in Scrubbers," Air Pollution Control, Part I (W. Strauss, ed.), Wiley-Interscience, New York, 1971.

Nukiyama, S., and Y. Tanasawa, "Experiment on Atomization of Liquid, Parts I and II," Trans. Soc. Mech. Engr. (Japan), 4:14, 86, 1938. Parts III and IV, 5, 63, 1939.

"Scrubbing System Removes Submicron Particulates," Chem. Engr., 78:21, 96, 98, September 20, 1971.

Semrau, K. T., "Dust Scrubber Design--A Critique on the State of the Art," J. Air Pollution Control Assoc., 13:12, 587-594, December 1963.

Stairmand, C. J., "The Design and Performance of Modern Gas-Cleaning Equipment," J. Inst. Fuels (London), 29, 58, February 1956.

Strauss, W., Industrial Gas Cleaning, Pergamon Press, London, 1966.

Chapter 22

ELECTROSTATIC PRECIPITATION

I. INTRODUCTION

Electrostatic precipitators (EsP), called electrofilters in
Europe and sometimes referred to as cottrells or electrical precipitators, have been used for more than 60 years (Cottrell, 1907) to
collect particulates, thereby preventing air pollution. The fact
that some of the early units are still in operation is testimony to
their durability when used in the proper applications. Many of
the EsP that were ill-designed for their tasks have not fared so
well--the most common complaint, that of failure to meet efficiency
specifications, has often resulted from shaving the size of the precipitator in order to meet competitive costs with other air cleaners,
but sometimes the failure resulted from a lack of information about
the size or resistivity of the particles to be collected. Despite
their high first cost, EsP are still finding frequent application as
the best buy for the collection of some dusts and mists. Properly
designed EsP will meet even the current stringent emissions regulations.

II. THEORY

The theory of operation for EsP is that particles in passing
through a corona discharge (ion field) will be charged with the
same polarity as the ions and may then be collected by an oppositely charged electrode. The usual industrial EsP has a relatively
small diameter discharge electrode which is given a large negative
potential (near the arcing voltage) with respect to the collecting
electrode, which is normally at ground, to produce a corona discharge (see Fig. 22-1). The corona discharge from the cathode

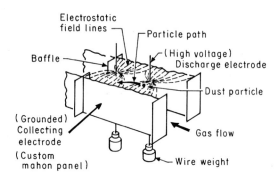

FIG. 22-1. Electrostatic precipitation. From "Mahon
Electrostatic Precipitation," Bulletin EP-68, R. C. Mahon, Detroit, Michigan, 1968.

gives a strong flux of electrons migrating toward the collecting
electrode (anode). Particles in passing through the electron flux
are negatively charged. The charged particles migrate, under the
influence of the electric field, toward the collecting electrode.
The particles adhere to the surface of the electrode until force is
applied to remove them or they are collected in a film of liquid
that irrigates the collecting surface and washes them off.

The exception to the usual industrial EsP uses a positive
field to charge the particles and they are collected on electrodes
that are negative with respect to the particles. This type of EsP
is usually a two-stage device with separate charging and collecting sections. The two-stage EsP are often called low-voltage EsP
because the charging field is much lower from the positive electrode
than from the negative electrode of the industrial EsP.

The theoretical migration velocity is derived from Newton's
law (mass times acceleration equals force) as

$$m \, d\omega/dt = q_c E_p - 6\pi\mu r \, \omega \,, \qquad\qquad (22\text{-}1)$$

where ω = migration velocity (cm/sec) = same as Ω in Chapter 5,
m = mass of particle (gm), q_c = charge on particle (esu), E_p = collecting field strength (esu), μ = dynamic viscosity of gas stream

(poise, gm/cm-sec), and r = particle radius (cm). The first term after the equal sign is the electrical force and the second term is the resistance to motion (see Chapter 5). Equation 22-1 integrates to

$$\omega = q_c E_p / (6 \pi \mu r) \, [1 - \exp(-6\pi \mu r t/m)] . \qquad (22\text{-}2)$$

Neglecting the exponential term (negligible for $t > \sim 10^{-2}$ sec) and substituting $q_c = n\epsilon = 3 E_c r^2 f$ plus including Cunningham's correction for small particle slip yield

$$\omega = [E_c E_p r / (2\pi\mu] f C_c , \qquad (22\text{-}3)$$

where E_c = charging field strength (esu), f = correction for dielectric costant $(\xi) = \xi/(\xi + 2)$, and C_c = Cunningham's correction = $1 + 0.86\lambda/r$, where λ = mean free path of gas molecule, see Chapter 5).

EXAMPLE 22-1: What is the migration velocity for a 1 µm-diameter particle from a wet process cement kiln if the ξ is 6.5,* the field strength is 4.5 kV/cm (54 kV with 10-in. spacing of collecting electrodes or about 12 cm clear field gap), and the temperature of the gas stream is 350°F?

[*See Table 22-1 for some typical values of ξ.]

Conversion of units in Eq. 22-3 gives the following:

$$\omega = 8.85 \times 10^{-5} E^2 df C_c / \mu \quad \text{with E in kV/cm and d in µm}$$
$$\omega = 8.85 \times 10^{-5} (4.5 \text{ kv/cm})^2 (1 \text{ µm}) \left[\frac{6.5}{6.5 + 2} \right] [1 + 1.7$$
$$\times (0.101/1)]/2.32 \times 10^{-4}$$
$$= 6.91 \text{ cm/sec or } 0.23 \text{ ft/sec}$$

[λ = 0.101 µm and µ = 2.32 $\times 10^{-4}$ poise (from Table 1-2)]

The actual ω may be only 1/3 to 1/2 of the theoretical value, primarily because the applied voltage fluctuates and the ω decreases as cake thickness increases. In practice the effective velocity for the total particle distribution may be used. Such a velocity is backcalculated from the efficiency obtained. Some values of effective velocities are given in Table 22-2.

III. DESIGN

The efficiency for EsP is a function of the driving force for

208   AIR POLLUTION

TABLE 22-1

Dielectric Constants[a]

Material	ξ	Material	ξ
Vacuum	1.0000	Phosphorus (red)	4.1
Air	1.0006	Potassium carbonate	5.6
Acetic acid	4.1	Potassium chloride	7.3
Alumina	4.5-8.4	Potassium nitrate	5.0
Ammonium chloride	7.0	Quartz	5.27-5.34
Calcium carbonate	6.14	Quartz (fused)	3.8-4.1
Calcium sulfate	5.66	Selenium	6.6
Cupric oxide	18.1	Sodium carbonate[b]	5.3
Dolomite	6.8-8.0	Sodium chloride	6.12
Ferrous oxide	14.2	Sodium nitrate	5.2
Glass	5-10	Sucrose	3.3
Glass (Pyrex)	3.8-6	Steam	1.01
Glycerine	56	Sulfur	4.0
Lead monoxide	25.9	Titanium dioxide	14-110
Lead sulfate	14.3	Urea	3.5
Lead sulfide	17.9	Water	80

[a]From various sources.
[b]With 10 H_2O.

TABLE 22-2

Effective Migration Velocities for Deutsch Equations[a]

Application	ω (ft/sec)
Pulverized coal fly ash	0.33-0.44
Paper mills	0.25
Open-hearth furnaces	0.19
Secondary blast furnace (80% foundry iron)	0.41
Gypsum	0.52-0.64
Hot phosphorus	0.09
Sulfuric acid mist	0.19-0.25
Flash roaster	0.25
Multiple-hearth roaster	0.26
Cement	
Wet process	0.33-0.37
Dry process	0.19-0.23
Catalyst dust	0.25
Gray iron cupola (iron:coke, 10:1)	0.10-0.12

[a]From J. A. Danielson, Ed., Air Pollution Engineering Manual,
National Center for Air Pollution Control Publ. No. 999-AP-40
Cincinnati, Ohio, 1967.

particle removal and the time during which the driving force acts. The driving force has a practical limit, dependent on the type of EsP and the characteristics of the particles and the gas stream. Therefore, the principal design variables are all related to the residence time of the particles in the collector.

A. Determination of Size

If laminar flow were maintained in the gas stream during collection, the minimum size for which 100% efficiency prevailed (d_{100}) would be that diameter of particle which had a migration velocity (ω) that would move the particle through the distance from the charging electrode to the collecting electrode (s) while the transport velocity (v) moved the particle through the length of the collector (L); i.e., $\eta = 1.00$ in

$$\eta = \omega L/sv , \qquad (22-4)$$

where η = fractional efficiency of collection for particles with migration velocity ω.

EXAMPLE 22-2: What is the length of the precipitator in Example 22-1 in order to obtain a theoretical 100% collection if v = 8 fps?

$\eta = 1.00 = 0.227$ fps X L/(8 fps X 12 cm)

L = 423 cm = 13.9 ft

[Note: A common design practice is to add 1 ft to the calculated length for a charging distance.]

Since the area of the collecting electrodes (A) is LHW /s and the flow rate (Q) is HWv where L, W, and H are the length, width, and height of the collector, Equation 22-4 may be written as

$$\eta = (A/Q) \omega . \qquad (22-5)$$

The gas flow in an EsP is not laminar; it is turbulent. The collection of particles approaches a random probability, and for a random probability the expression for efficiency becomes the familiar Deutsch equation,

$$\eta = 1 - \exp(-A\omega/Q) . \qquad (22-6)$$

For analogy with stirred settling, see Chapter 5. The exponential
term in Eq. 22-6 is the penetration (q) as previously used. After
A/Q is fixed, Eq. 22-6 may be rewritten as

$$\eta = 1 - \exp(-kd) , \qquad\qquad (22\text{-}7)$$

where $k = 0.029\,(A/Q)\,E^2 f C_c/\mu$ (μm^{-1}), A in ft^2, Q in ft^3/min,
and d in μm.

EXAMPLE 22-3: What collection efficiency should be expected
    on the overall dust in Example 22-1, according to the
    Deutsch equation, if the length of the precipitator is 30 ft?

Area of collecting electrodes:

$$A = 30\,HW/(12/30.48) = 76.2\,WH \; ft^2$$

Volumetric flow rate:

$$Q = W\dot{H}v = WH(8 \times 60) = 480\,WH \; ft^3/min = 8\,WH \; ft^3/sec$$

Effective migration velocity = 0.35 ft/sec (from Table 22-2)

$$\eta = 1.000 - \exp[-76.2\,WH \times 0.35/(8\,WH)] = 1.000 - \exp(-3.33)$$
$$= 0.964 \text{ or } 96.4\%_w$$

[Note: For 90,000 cfm or about 2000 bbl/day capacity,
    $A = 13,500$ ft^2.]

Electrode spacing (2 s between collecting electrodes) is
usually set between 6 and 15 in., with an average of about 10 in.
For low-voltage precipitators, the spacing may be as small as
1/2 in. The spacing should be sufficiently large that the thickness
of two collected dust layers will not seriously affect the operation
between the electrodes, yet small enough to get large corona power
leakage at reasonable operating voltages. Charging electrodes
are spaced along the air path between collecting electrodes at
intervals of 4 to 15 in. Tubular collecting electrodes are normally
smaller in diameter than the spacing of the plate electrodes.

The gas velocity needs to be low to prevent scour of the
collected material and high to get reasonable width and height of
the EsP unit. Maximum gas velocities permitted by various applications range from 3 to 20 fps; reentrainment usually starts at

about 3 fps for flocculent, conductive particles (carbon black),
8 fps for fly ash, 10 to 12 fps for cement kiln dust, and 15 to 20
fps for mists or irrigated electrodes. There has been a recent
reversal of the trend toward higher velocities with 5 to 6 fps being
used where 8 fps would have been used a few years ago. The precipitators are being made this much larger in the interest of better
collection. The width employed in the calculation of the velocity
should be the net width after subtracting the thickness of electrodes
and average dust cakes.

The width and height combination of the EsP are determined
from the transport velocity through the unit and the flow rate. The
width and height are of the same order of size, rarely differing by
a factor 2. The length of the EsP is then determined by the penetration or efficiency to be obtained. It is highly desirable to have
multiple sections along the length of the EsP so that the latter
sections can collect the dust which is reentrained in the early
sections during rapping. It is desirable to have multiple rows of
sections along the width of the EsP for continued operation when
a section must be removed from operation.

B. Types of Electrodes

Electrodes are made in many different shapes for the stated
purpose of obtaining the best efficiency at the lowest cost. The
charging electrode shape is designed to maximize the corona
leakage (small diameter or shapes with sharp corners or protrusions
such as twisted square or barbed wires) and the collecting electrode shape is made to avoid scour (see Fig. 22-2). It is doubted
by many engineers in the field that a shaped discharge electrode
is better than a smooth wire (the nonround shapes are more popular in Europe than in the United States) and some even doubt the
efficacy of shaping the collecting electrode rather than using a
flat plate. One manufacturer uses perforated plates for the collecting electrodes and has the air flow perpendicularly through

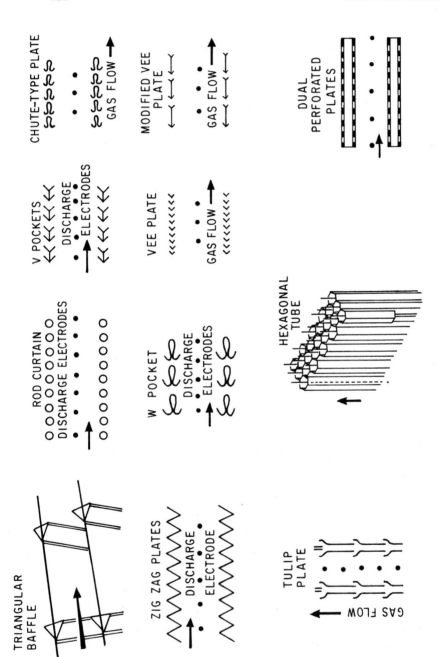

FIG. 22-2. Collecting electrode shapes.

the collecting electrodes rather than parallel to them. Objections
to the complex shapes in electrodes are that the removal of the
collected material is difficult, the spacing is harder to maintain
for the complex shapes than it is for the simple ones, and the
cost of shaped electrodes is high.

C. Operating Parameters

In order to assure success with EsP, it is necessary that
careful attention be given not only to the maintenance of the
hardware and the design specifications but also to the operation
of the EsP. The operating voltage, particle characteristics, and
electrode cleaning require consideration.

1. Operating Voltage

The operating voltage should be kept as high as possible
for the maximum corona leakage. The minimum penetration is
obtained with some sparking, despite the fact that sparking knocks
holes in the collected dust layer and reentrains some dust. Sparkover depends on the operating voltage, electrode spacing, types
of electrodes, gas composition, particle concentration, dust layer,
and bulk resistivity of the dust. Automatic controls lower the
voltage when sparkover occurs to prevent continued arcing.

The operating voltage is rectified ac. Older installations
are likely to be half-wave and the newer ones full-wave rectifi-
cation. Silicon rectifiers, because of their higher efficiencies,
are replacing many mechanical and electronic rectifiers. Current
research is aimed at a square wave voltage. The peak voltages
for industrial high-voltage EsP are mainly 30 to 75 kV but may be
as high as 100 kV. The field strengths will usually be 4 to 5 kV/cm.
The voltage for two-stage or low-voltage EsP will normally be about
12 to 13 kV in the charging section and about half that in the collector.

2. Corona Power

The 30- to 75-kV peak voltage and the 50- to 600-mA current

per section (0.1 to 1 mA/ft of wire) give a corona power of 50 to
500 W/1000 cfm. Corona power is related to the penetration and
efficiency by

$$\log q = \log(1 - \eta) = KP/Q , \qquad (22-8)$$

where q = fractional penetration, P = corona power level (W),
Q = flow rate (cfm), and K = empirical constant (-0.02 to -0.2
cfm/W).

3. Bulk Resistivity and Dielectric Constant

The bulk resistivity and the dielectric constant for the
particles to be collected will determine to a large extent the
efficiency of collection.

The bulk resistivity (ρ_B, ohm-cm) can be measured by placing
a dust sample of known area (A, cm^2) between two electrodes and
measuring the voltage (V), current (I, amps), and distance (x, cm)
between the electrodes:

$$\rho_B = AV/Ix . \qquad (22-9)$$

ρ_B is a function of temperature, moisture, and chemical makeup
of the particles (see Figs. 22-3). ρ_B must be in the range of
10^4 to 2×10^{10} ohm-cm for satisfactory EsP collection. If ρ_B is
too low (carbon black), the collected particles take on the charge
of the collecting electrode and become reentrained; if ρ_B is too
high (dry limestone), back corona occurs because of the large
potential drop across the collected material.

The dielectric constant (ξ) enters into the migration velocity
of particles because it affects the amount of charge that particles
will take. The resistance of dielectrics (glass, ceramics, etc.)
decreases with increasing temperature.

4. Removing Collected Material

The particles collected must be removed from the collecting

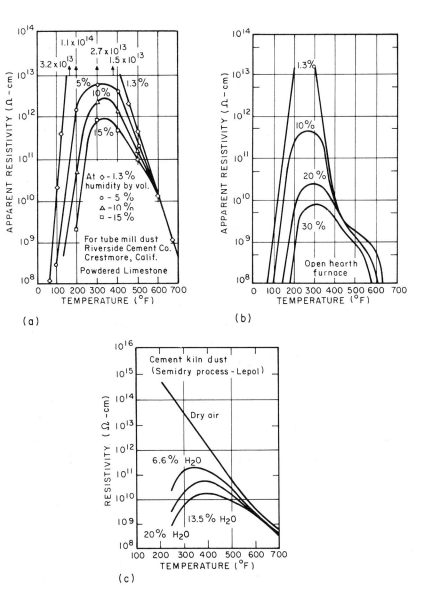

FIG. 22-3. Apparent resistivities of particles. (a and b) From W. T. Sproull and Y. Nakada, "Operation of Cottrell Precipitators," J. Ind. Engr. Chem., 43, 1350, 1951. (c and d) From H. J. White, Industrial Electrostatic Precipitation, Addison-Wesley, Reading, Massachusetts, 1963.

electrode. In low-voltage EsP applications, the buildup is slow enough that the cleaning is normally done by shut down of the unit; however, for the industrial high-voltage EsP, the dust buildup is rapid and removal must be done often and usually while the unit is in service.

Probably the most widely used removal mechanism is the rapping of the electrodes. Hammers are used to tap the electrode hangers and cause the dust layer to fall off. The maximum dust layer should be relatively thick (about 1/2 in.) for most efficient operation; it sometimes sloughs unevenly or slides only part way down during rapping. How hard to rap and when to rap are vital questions that must be carefully answered in good design. The best EsP design will provide for variations in the rapping parameters so that the actual procedures can be developed by operating experience. Magnetic impulse rapping or compressed air rapping is often done at 1 to 2 raps per minute per electrode for the first sections. Vibrating the electrodes at about 50 to 100 Hz may be used for cleaning. Cleaning the electrodes, especially by rapping, releases puffs of dust. As a result, different portions of the EsP should be rapped at different times. It is important to have multiple sections of the EsP in the direction of gas flow so that the electrode cleaning may be sequenced. The first section should be cleaned relatively often and the last section much less frequently.

Mists run off the collecting electrode as they are collected. Irrigated precipitators spray a liquid (usually water but sometimes oil) over the collecting electrode to wash the collected material from the electrode. Irrigation of electrodes is not widely practiced because it removes one of the principal advantages of EsP, namely, that of dry collection. However, irrigation may permit the application of EsP to sticky particles such as hydrated lime, tar, or grease where the only alternative is wet scrubbing anyway. One major equipment manufacturer and its client learned the hard way that brackish bay water should not be used for irrigation of the electrodes, which seems almost obvious if any consideration at all is given to the possibility of corrosion.

Corona electrodes build up ring-shaped (donut) deposits of dusts; therefore, the charging electrodes also require cleaning. The cleaning is usually done by rapping or vibrating, but may be done by manual rapping on the premise that it is not often required.

IV. APPLICATIONS

The first requisite for satisfactory application of the EsP is that the aerosol have the proper resistivity. If the bulk resistivity of solid particles is not in the range of 10^4 to 2×10^{10} ohm-cm, EsP can be used only with design changes in the equipment or the process. Low resistivity dusts tend to be released and reentrained in the gas stream; they require low velocities or irrigation. High resistivity materials are not readily charged for collection and once collected they cause the voltage drop to be across the dust layer and lead to back corona; they require adjusting the resistivity by change in temperature or by the addition of water, sulfur trioxide to basic dust, ammonia to acid dust, or other material (see Chapter 13).

Almost every fly ash has enough high resistivity components to preclude its collection at very high efficiency $(99.5\%_w)$ without lowering the ρ_B. This is the reason for the upsurge in popularity of hot-side precipitation to meet the current emission limits. As can be noted in Fig. 22-2, cleaning the gas stream at 700°-800^oF offers much improvement over the 250°-300^oF for the usual exit temperature. The volume of gas to be cleaned will be about 60% greater for the hot-side precipitator than for the usual application temperature. Hot-side precipitation gives the additional advantage of less plugging of the secondary heat exchangers.

The advantages of high-voltage EsP include low head loss, dry collection (for return to process and collection without corrosion problems), high temperature capability, abrasive and sticky materials collection, collection of mists without dilution, high efficiency in proper usage, long life, and minimum maintenance. The disadvantages of EsP include high first cost, noise, high-voltage dangers to personnel and for fire and explosion, ozone formation, ρ_B limitations, large space requirements (2- to 10-second residence times), and care required in maintenance. Low-voltage EsP has low efficiency for its size and it is difficult to clean, but the fact that low voltage is inefficient in ozone generation gives the low-voltage EsP one distinct advantage over high-voltage EsP.

High-voltage EsP units are widely used for fly ash from power plants, acid mists from acid manufacturing, abrasive dusts from metals industries, fumes from smelters (especially smelters other than lead), cement kiln dust, coal and lignite drying, coke ovens, pyrites roasters in paper, chemical, and smelting industries, alumina calciner, and many other applications.

Two-stage (low-voltage) EsP have been used to collect oil mists from machining, tar and asphalt particles from saturators (roofing manufacturing), emissions from meat smokehouses, and particulates from any air that is to be breathed soon after cleaning, such as the air inside submarines.

V. COSTS

Costs of EsP vary widely with application and penetration required. The low emission limits that are being set today preclude the 90-95%$_w$ efficiencies of the recent past and often make an efficiency of 99.5%$_w$ mandatory. The cost difference is reflected in the number of transfer units (2.3 for q = 0.10 versus 5.3 for q = 0.005), and the purchase cost estimates for many applications have risen from about $1/acfm to $2-6/acfm. The acfm at power plants will nearly double for hot-side precipitation. For additional cost information, see Section IV, pp. 152-154.

EXAMPLE 22-4: Estimate the total annual cost for a high efficiency, 100,000-acfm electrostatic precipitator with a purchase cost of \$3/acfm.

Capital costs: 10 yr amortization at 10%

Installed cost = 225% of first cost

$$C_a = \frac{P}{N} \frac{(1 + R)^m}{(1 + R)^n - 1}$$

$$= [2.25(3)(100,000)/10](1 + 0.10)^{10}/[(1 + 0.10)^{10} - 1]$$

$$= \$109,853.14$$

Power costs:

Corona power: 400W/1000 acfm (100,000 acfm) = 40 kW

Head loss: 1.2 in. w.g.

$$kW = \frac{0.746(100,000)(1.2)}{6356(0.60)} = 23.5$$

Total: (40 + 23.5) kW X 8760 hr/yr X \$0.015/kWh

$$= \$8340.47$$

Maintenance: \$0.032/acfm (100,000 acfm) = \$3200

Total cost: \$109,850 + \$8340 + \$3200 = \$121,400 per year

VI. SUMMARY

Particles suspended in a gas stream may be removed by electrostatic precipitation. The process consists of passing the aerosol stream through a strong electric field (typically 4-5 kV/cm) where the particles receive a charge from the corona discharge, then collecting the charged particles on oppositely charged electrodes.

Electrostatic precipitators can satisfactorily remove solid particles with bulk resistivities in the range of 10^4 to 2×10^{10} ohm-cm. The method works well on even highly conductive liquid particles. The units are quite large in order to provide residence times up to several seconds. The collected material is usually removed from the electrodes by rapping. The mists drain from collecting electrodes, and irrigation is used in some applications to wash solids from the electrodes to prevent reentrainment or sticking of the collected material.

Electrostatic precipitation is the cleaning method of choice for power plant fly ash and many smelting fumes. High efficiency cleaning by this method can be achieved at a total annual cost which is generally lower than that for high-energy scrubbing but higher than that for bag filtration (see Fig. IV, p. 154).

PROBLEMS

1. What is the force in g's being applied in Example 22-1? Ratio of ω to settling velocity for that size particle with a density of 2.7 gm/cm^3.

2. Plot a graph (log X log) of the force (g) versus particle diameter (μm) for the conditions of Example 22-1.

3. Plot particle velocity versus migration force for particles with diameters of 0.5 μm, 1 μm, and 10 μm. Do the plots deviate from straight lines? Why?

4. On the plot made in Problem 3, draw envelope curves for the areas usually covered and possibly covered by EsP designs.

5. Estimate the total annual cost of an EsP to clean 80,000 acfm if the purchase cost is $3.25/acfm, amortization is 15 years at 10%, head loss is 1.6 in. w.g., corona power is 390 W/1000 acfm, and the cost of electricity is $0.020/kWh.

6. Plot efficiency versus diameter (log X arith) for k = 0.1, 0.2, 0.5, 1, 2, 5, and 10 μm^{-1} according to Eq. 22-7.

7. Determine the probable dimensions for an EsP to clean 4×10^6 acfm if a residence time of 8 sec is required to obtain the penetration limit. How many sections should be made in this EsP?

8. Calculate the collecting electrode area and the effective migration velocity for Problem 7 if electrode spacing (2s) is 10 in. and penetration (q) is 0.005.

BIBLIOGRAPHY

Baxter, W. A., "Recent Electrostatic Precipitator Experience with Ammonia Conditioning of Power Boiler Flue Gases," J. Air Pollution Control Assoc., 18:12, 817-820, December 1968.

"Big Powerplant Precipitator," Air & Water News, 5:33, 4, August 23, 1971.

Busby, H. G., and K. Darby, "Efficiency of Electrostatic Precipitators as affected by the Properties and Combustion of Coal," J. Inst. Fuel, 36, 184-197, 1963.

Crawford, W. D., "The Cost of Clean Energy," J. Air Pollution Control Assoc., 19:5, 322-324, May 1969.

Danielson, J. A., Ed., Air Pollution Engineering Manual, National Center for Air Pollution Control Publ. No. 999-AP-40, Cincinnati, Ohio, 1967.

"Electrostatic Precipitators: New Designs Spell Wider Use," Chem. Engr., 77:16, 32-33, July 13, 1970.

"Hot, Hot Precipitator," Air & Water News, 6:2, 5, January 17, 1972.

Katz, J., "The Effective Collection of Fly Ash at Pulverized Coal-Fired Plants," J. Air Pollution Control Assoc., 15:11, 525-528, November 1965.

McLaughlin, J. F., "Air Pollution Control in Coal Fired Steam-Electric Plants," presented at Air and Water Pollution Conference, Columbia, Missouri, November 19, 1957.

O'Connor, J. R., and J. F. Citarella, "An Air Pollution Control Cost Study of the Steam-Electric Power Generating Industry," J. Air Pollution Control Assoc., 20:5, 283-288, May 1970.

Oglesby, S., Jr., "Electrostatic Precipitators Tackle Air Pollutants," Environmental Sci. & Tech., 5:9, 766-770, September 1971.

Penney, G. W., "Some Problems in the Application of the Deutsch Equation to Industrial Electrostatic Precipitation," J. Air Pollution Control Assoc., 19:8, 596-600, August 1969.

Reese, J. T., and J. Greco, "Experience with Electrostatic Fly-Ash Collection Equipment Serving Steam-Electric Generating Plants," J. Air Pollution Control Assoc., 18:8, 523-528, August 1968.

Robinson, M., "Electrostatic Precipitation," Air Pollution Control, Part I (W. Strauss, ed.), Wiley-Interscience, New York, 1971.

Schrader, K., "Improvement of the Efficiency of Electrostatic Precipitation by Injecting SO_3 into the Flue Gas," Combustion, 42:4, 22-28, October 1970.

Strauss, W., Industrial Gas Cleaning, Pergamon Press, New York, 1966.

White, H. J., Industrial Electrostatic Precipitation, Addison-Wesley, Reading, Massachusetts, 1963.

Chapter 23

AIR-CLEANER COMBINATIONS AND OTHER METHODS

I. INTRODUCTION

The air-cleaning methods that have been described in the preceding chapters are sometimes used in combination. Of course, it is the usual practice to have the venturi scrubber and the louver concentrator followed by cyclones. In addition, there are some other cleaning and/or conditioning processes that have been or may be occasionally used.

II. AIR-CLEANER COMBINATIONS

Combinations of air-cleaning methods are used for removing more than one type of pollutant or for precleaning to prevent clogging of a high efficiency or catalytic or adsorption process. A mechanical cleaner may enhance the efficiency of a following electrostatic precipitator if it takes out some particles that would not be collected because of their resistivity. A conditioning system, especially for particle size enhancement, can be used in conjunction with cleaners whose efficiencies are size dependent.

A. Precleaning of Components

A component of a waste gas stream can be removed ahead of the primary cleaning process for the purpose of protecting the cleaner. For example, particulate collection may be used before a catalytic unit or an activated carbon bed or even a packed tower when the amount of particulate present would not otherwise justify cleaning for air pollution control purposes. Prefilters or roughing filters such as fiberglass beds have been used ahead of HEPA

filters. Also, a mist eliminator is usually included in the supply
line to a flare.

Two or more air cleaners in series to remove multiple pollutants can be desirable, although some air cleaners are capable
of removing more than one type of pollutant; for example, the
venturi scrubber and other scrubbers can remove both gaseous and
particulate pollutants. The exit gas from an afterburner may require cleaning of particulates and/or gases, ashes or dusts and
obnoxious or toxic gases such as sulfur dioxide, nitrogen oxides,
oxides of phosphorus, hydrogen chloride, and others.

B. Particle Size Enhancement

Particles that collide tend to stay together as a single,
larger particle. Although Brownian motion will cause particles
to coagulate, the rate of such coagulation is far too slow to be
practicable. Turbulent motion of the aerosol also causes differential velocities that improve agglomeration, but turbulent coagulation is still too slow. The most logical way to increase the
collision frequency is by using sound to increase the motion of
the particles. Sonic agglomeration has been used with sulfuric
acid mists, carbon black, lead oxide, zinc oxide, and other
small aerosols. Good efficiencies require particle loadings well
above 2 gr/Ncf, sound levels over 150 dB (0.1 W/cm^2), and agglomeration times up to 4 sec. Time required at 165 dB may be
on the order of 2 sec. The usual frequency is a few kilohertz.
The differential velocities for different diameter particles result
in agglomeration.

For sonic agglomeration, the dirty gas stream is usually
put through a column with the whistle or siren located at one end,
then a multiple cyclone. According to Strauss, the installed cost
for a sonic agglomeration cleaner is somewhat less than for an
electrostatic precipitator. The rather high maintenance costs
together with the noise problems created have tended to discourage
the use of sonic agglomeration. Sonic agglomeration of fogs at

airports has been tried; the idea was abandoned because in the absence of standing waves, sound intensities required for agglomeration are intolerable.

Condensation of water onto small particles can increase particle size. This phenomenon probably plays a minor role in the venturi scrubber. However, there are two major drawbacks to application of condensation as the size enhancement process: it is difficult to maintain a supersaturated condition for condensation to occur and the amount of water required to build up the size of the particles when a droplet contains only one particle is tremendous. For example, 10^6 volumes of water would be required per volume of particles to increase particle size from 0.1 to 10 μm. Where diffusiophoresis and/or Brownian impaction complement the condensation to give more than one particle in a droplet, the process becomes more nearly a possible practical application.

C. Other Reasons for Series Cleaning

Putting air cleaners in series to increase the collection efficiency has often been done. Another section of electrostatic precipitator may be added to an existing precipitator; one or more additional cyclones may be added in series to a catalyst recovery operation; a cyclone may be used in conjunction with an electrostatic precipitator for fly ash collection; and an additional absorbing section may be added to lower the penetration of sulfur dioxide through an acid plant.

Series operation can serve purposes other than increased efficiency. These purposes include improved versatility of operation, classification of collected materials, and fire protection. Particles may be sorted by size or density in a series of collectors, and gaseous materials may be separated by solubility or condensation temperature when collected in series. The plugging of the dust discharge from a cyclone used for catalyst recovery will not shut down the operation if the following cyclone takes over the added load; this advantage adds to that of increased collection by the series operation. Combustibles may be collected separately when they have different collection characteristics. Large cenospheres may need to be collected separately because of their longer incandescent state than the smaller particles.

III. OTHER METHODS

Any of the forces which result in the movement of aerosols could theoretically be used for particle collection. In addition to the previously described movement forces, these include migration from thermal forces, photon bombardment, and magnetic deflection. Of these, only thermal migration appears practically possible. Although thermal precipitation is used for particle sampling, it has not proved practicable on an air-cleaning scale. There have been patents based on the principle; however, it has not been feasible to maintain the temperature gradients necessary for application. The apparent concern with fine particles will undoubtedly cause new evaluations of thermal precipitation because the process does have very small particle capabilities.

Under certain conditions many of the processes which are not usually practical may be the best methods to use; therefore, the unusual should not be rejected out of hand without giving the possibility of use some consideration. Furthermore, there is the concerted effort by some people in the field of air pollution control to require the collection of even the very small aerosols. Such a requirement may very well be instituted because of the present methods of proving detrimental health effects; putting in this requirement could change the feasibility status of some of the processes that are strictly theoretical now.

IV. SUMMARY

Application conditions may warrant the use of combinations of air cleaners or of unusual air-cleaning schemes. Many innovative schemes for particle collection founder on attempts to violate the basic principle involved in the method. For example, the distance for accelerating or decelerating a 1-μm particle to or from 5000 cm/sec in air is less than a mm; some of the power-saving collectors that are being introduced would require this distance to be several cm in order for them to work.

The normal combination of cleaners is the use of a precleaner to prevent fouling a catalyst, an adsorption bed, or an HEPA filter. Series cleaning has been used in catalyst recovery and some other operations in the interest of improving efficiency. A series of inefficient cleaners is not likely to meet the stringent air pollution regulations now in effect.

Sonic agglomeration has been used to build up particle size for easier collection. The process has not found wide application because of high maintenance and operating costs and the difficulty of containing the noise. Condensation of water onto small particles to enhance particle size shows theoretical possibilities and advantages; practical application remains to be developed.

Thermal precipitation and other particle migration principles from various external forces can collect particles, even very small particles, but so far not in practical application.

PROBLEMS

1. How many cyclones in series can pay their own way if the flow rate is 10,000 acfm, the particle loading is 100 gr/ft^3, the first unit collects 84%$_w$, the second unit and following units will be 30% as efficient on the material reaching them as the preceding unit, and the value of the material is $80/ton?

2. What are the relative merits of trying to collect fly ash from coal burning at 99.5%$_w$ with a combination of a multiple cyclone and a cold side electrostatic precipitator (300°F) versus moving the electrostatic precipitator to the hot side (740°F) and using it alone?

3. Design a collector which uses 4-in.-wide ribbons stretched
   from top to bottom of the collection box for diffusional
   removal of 0.1-μm particles with a density of 5 gm/cm^3
   from 10,000 acfm of gas at 20°C.

4. What are the advantages of a collector such as described in
   Problem 3? What are its disadvantages?

BIBLIOGRAPHY

Junge, C. E., "Methods of Artificial Fog Dispersal and Their
    Evaluation," Air Force Surveys in Geophysics, No. 105,
    September 1958.

St. Clair, H. W., "Agglomeration of Smoke, Fog, or Dust Particles
    by Sonic Waves," Ind. Engr. Chem., 41:3, 2434-2438, 1949.

Stern, A. C., Ed., "Efficiency, Application, and Selection of
    Collectors," Air Pollution, Vol. III, 2nd ed., Chap. 42,
    Academic Press, New York, 1968.

Strauss, W., Industrial Gas Cleaning, Pergamon Press, New
    York, 1966.

Section V: Specific Controls

Air pollution control usually requires a synthesis of the information on a number of topics in order to solve a single problem. Such an approach is the rationale used here on both specific pollutants and specific sources.

Chapter 24

SPECIFIC POLLUTION CONTROLS

I. INTRODUCTION

Controls have prevented air pollution problems from the release of particulate pollutants, odors, and toxic substances on a widespread basis for a relatively long time. For pollutants such as sulfur dioxide, the practice has generally been problem prevention through dilution; however, the pressures against the dilution of materials by spreading through the environment have increased with population density and with an increased awareness of and concern about all forms of pollution. Regulations that are being set usually prescribe emission limits which challenge our removal technology, especially within the realm of feasibility.

II. SULFUR DIOXIDE

Sulfur dioxide control can be justified on the basis of esthetic insult (odor and visibility) and vegetation damage without invoking the tenuous and debatable health effects that have been attributed to even low concentrations of sulfur dioxide. The principal sulfur dioxide controls are dilution by tall stacks, use of low sulfur fuels, and cleaning of flue gases. Federal standards for new power plants limit sulfur dioxide emissions to 1.2 lb/10^6 Btu fired (equivalent to about 1.5%$_w$ sulfur) for solid fossil fuel plants and 0.8 lb/10^6 Btu fired (equivalent to about 2%$_w$ sulfur) for oil-fired plants. These regulations specify a limit of 4 lb of sulfur dioxide per ton of sulfuric acid produced for acid plants.

A. Dilution

Atmospheric dilution has prevented many acute problems

231

with sulfur dioxide. Stacks up to 1250 ft in height have been erected for smelter gases and to 1206 ft for power plants. Dilution will continue to serve a useful purpose until direct control methods are feasible. During adverse meteorological conditions, the dilution has been accompanied by the switching of fuels at power plants and the curtailment of operations at smelters.

B. Low Sulfur Fuels

Scrubbing hydrogen sulfide from sour natural gas with monoethanolamine or similar solvent has long been common practice. The resulting hydrogen sulfide has often been flared and, when sulfur recovery was practiced, it was usually done by the Claus process; however, the more efficient Stretford process is coming into common usage to meet new regulations. The Claus process is a dry catalytic conversion process which produces molten sulfur; the Stretford process converts the sulfur to elemental form in a wet catalyst-liquid flotation method.

Desulfurization of oil has become routine in the last several years. The removal is done at the refinery and costs about $0.65 per barrel. The sulfur from the oil has added to that from the sour gas to glut the sulfur market and depress the price of sulfur, a situation that could possibly be changed with development of new uses for the sulfur.

Low sulfur coal has been obtained by selective mining and by washing or treating the coal to remove the sulfur; neither method is very satisfactory. Coal gasification appears to be the method that will be chosen for using our vast coal reserves without causing major sulfur dioxide pollution problems. The coal gas is scrubbed of sulfur and other unwanted materials and converted to methane to achieve a high Btu content. The conversion to methane is difficult but highly desirable for permitting the continued use of present gas-burning equipment and for economical transmission over long distances.

C. Removal Processes

There are more than 50 sulfur dioxide removal processes
under various stages of development; many of them are subsidized
by the Environmental Protection Agency in its quest for the best
way to remove relatively dilute sulfur dioxide from flue gases.
Although there are several acid plants on smelter gases, there are
only a few sulfur dioxide removal units as large as 100 MW on
power plant gases (see Table 24-1). When wet scrubbing is used
for sulfur dioxide removal, care must be taken that higher ground
concentrations do not result after removal at usual efficiencies of
less than 70% than occurred without removal; the plume loses its
buoyancy in the process. The discussion that follows is limited
to brief descriptions for several of the sulfur dioxide removal
processes that are most prominent in current literature.

1. Limestone or Lime Scrubbing (Throwaway)

Pulverized limestone injected into the furnace reacts with
the sulfur dioxide to form a gypsum slag. The TVA found this
dry-scrubbing process took out about half of the sulfur dioxide
at a cost of over $2/ton of coal versus an estimated $0.75 and
increased the amount of particulate escaping from the electrostatic precipitators. Combustion Engineering (C-E) removed the
resulting particulates, fly ash, and some of the unreacted sulfur
dioxide with marble bed scrubbers at St. Louis (Union Electric's
Meramec Station) and at Lawrence, Kansas (Kansas Power and
Light). These installations which were started in 1968 have had
major problems with plugging in reheaters and scrubbers along
with low efficiency of removal and high additive requirements.
The St. Louis operation shut down in September 1972 because of
operational difficulties.

Scrubbing sulfur dioxide with an alkaline (limestone) water
is being studied by C-E and others. The Battersea process uses
about 8000 gpm of alkaline Thames River water with 33 lb/min of

TABLE 24-1

Sulfur Dioxide Removal at Fossil Fuel Power Plants[a]

Utility company/plant	Size (MW)	Scheduled startup	Fuel/% S
Limestone scrubbing			
Union Electric Co. (St. Louis)/ Meramec No. 2	140	September 1968[b]	Coal/3.0
Union Electric Co. (St. Louis)/ Meramec No. 1	125	Spring 1973	Coal/3.0
Kansas Power & Light/Lawrence Station No. 4	125	December 1968	Coal/3.5
Kansas Power & Light/Lawrence Station No. 5	430	November 1971	Coal/3.5
Kansas City Power & Light/Hawthorne Station No. 3	130	Mid-1972	Coal/3.5
Kansas City Power & Light/Hawthorne Station No. 4	140	Mid-1972	Coal/3.5
Kansas City Power & Light/La Cygne Station	820	Late 1972	Coal/5.2
Detroit Edison Co./St. Clair Station No. 3	180	November 1972	Coal/2.5-4.5
Detroit Edison Co./River Rouge Station No. 1	270	December 1972	Coal/3-4
Commonwealth Edison/Will Co. (Chicago) No. 1	175	February 1972	Coal/3.5
Northern States (Minnesota)/Surban Co. Stations 1, 2	680	May 1976	Coal/0.8
Arizona Public Service Co./Cholla Station	115	January 1973	Coal/0.4-1
Tennessee Valley Authority/Widow's Ck. Sta. No. 8	550	April 1975	Coal/3.7
Duquesne Light Co. (Pittsburgh)/Phillips Station	100	February 1973	Coal/2.3
Louisville Gas & Electric Co./Paddy's Run No. 6	70	Late 1972	Coal/3.0
City of Key West (Florida)/Stock Island[c]	37	June 1972	Oil/2.75
Sodium-based scrubbing			
Nevada Power Co./Reid Gardner Station	250	Mid-1973	Coal/1.0
Magnesium oxide scrubbing			
Boston Edison/Mystic Station No. 6[c]	150	March 1972	Oil/2.5
Potomac Electric & Power (Maryland)/Dickerson No. 3	195	Early 1974	Coal/3.0
Catalytic oxidation			
Illinois Power/Wood River[c]	100	June 1972	Coal/3.5

[a]From J. Air Pollution Control Assoc., 22:6, 474, June 1972.
[b]Discontinued operation September 1972.
[c]Partially funded by the Environmental Protection Agency.

$CaCO_3$ (chalk) to remove about 20 lb/min of sulfur dioxide (~90% efficiency) from 330,000 Ncfm (dry) of flue gas. The scrubbing liquor is settled and oxidized to convert sulfides to sulfates; then it is mixed with 90,000 gpm of cooling water and returned to the river.

Limestone scrubbing is the method of choice for 16 of the 20 operating, or planned for the near future, sulfur dioxide removal systems at power plants. The only existing units operating without subsidization are in this category. A venturi scrubber with lime slurry can remove up to about 70% of the sulfur dioxide while removing the fly ash as well. A scrubber put in for fly ash removal now and sulfur dioxide removal later has had severe corrosion problems.

2. Sulfuric Acid Production

Some components of smelter gases have sulfur dioxide concentrations high enough (up to several $\%_v$) to make acid production feasible; however, the dilute gases from power plants ($\leq$ few tenths $\%_v$) cannot economically produce acid. The methods for the production of acid from stack gases are catalytic (V_2O_5) oxidation after particulate removal (Monsanto) and removal and catalysis by activated carbon (Westvaco, Hitachi, and Cherniebau). The acid produced by these processes is usually weak acid, which has very poor marketability.

Acid is also produced in a magnesium oxide-scrubbing (Chemico/Basic) process that has been put on a 150-MW plant in Boston. The magnesium sulfide produced in scrubbing at 200°- 300°F is calcined at 1400°F to give magnesium oxide for scrubbing and sulfur dioxide that is put into contact process acid production.

3. Ammonia and Other Scrubbing

Ammonia scrubbing of sulfur dioxide at the smelter in Trail, British Columbia (COMINCO) produces ammonium sulfite which is

treated with sulfuric acid to get ammonium sulfate and sulfur
dioxide. The sulfur dioxide goes to acid production. The fertilizer market for the ammonium sulfate has been severely reduced
by the use of anhydrous ammonia as the nitrogen source.

A venturi scrubber with sodium carbonate has been designed
to remove 90% of the sulfur dioxide from the flue gas of a plant
burning $1\%_w$ sulfur coal (Combustion Equipment Associates for
Nevada Power). The sodium carbonate liquor gives a higher efficiency for the short contact period than the lime and limestone
slurries previously mentioned.

Other sulfur dioxide-scrubbing processes have been developed and applied on a limited scale, especially to smelter
gases. The Sulphidine Process (Lurgi) uses a 1:1 mixture of
xylidine and water as the absorbing liquor. Dimethylaniline can
be used to scrub out the sulfur dioxide (ASARCO). Apparently,
other organic solvents are being developed as absorbing liquors
(NOSOX by Monsanto). The sulfur dioxide is stripped from the
absorbent and sent into acid production by conventional plants.

III. OXIDES OF NITROGEN

Nitric oxide is produced in concentrations up to a few
thousand ppm_v in the flue gases from combustion processes.
The total amounts of nitrogen oxides formed by combustion are
less than those produced by nature, but the concentrations within
small areas are greater from the combustion processes. Federal
regulations for new stationary sources limit the emissions of
nitrogen oxides from power plants to 0.2 $lb/10^6$ Btu input for
gas fired, 0.3 $lb/10^6$ Btu input for oil fired, and 0.7 $lb/10^6$ Btu
input for coal fired. They also limit the nitrogen oxides (as NO_2)
to 3 lb/ton of acid produced at nitric acid plants. Only recently
have we become interested in controlling nitrogen oxide emissions;
the preliminary results of control efforts are quite promising.

A.   Combustion Modifications

Nitrogen oxides are formed in combustion processes according to

$$N_2 + O_2 \xrightarrow[k_2]{k_1} 2\,NO \qquad\qquad (24\text{-}1)$$

and

$$2\,NO + O_2 \xrightarrow[k_4]{k_3} 2\,NO_2 \rightleftharpoons N_2O_4 \,. \qquad\qquad (24\text{-}2)$$

The reaction equilibrium constants (k_i) depend on the concentrations of the reactants and products, but more strongly on the temperature and the cooling rate of the gases, and to a lesser extent on the firing chamber wall catalysis. The kinetics equations that have been devised do not predict the formation of nitrogen oxides very well; they show that the amount of NO_x is proportional to the concentration of nitrogen, but recent experiments with argon substituted for the nitrogen in the combustion gas indicate that the organic nitrogen in the fuel is usually sufficient to cause the normal amount of nitric oxide. Organic nitrogen is more available than gaseous nitrogen. Bartok et al. correlated the amount of nitrogen oxides with the size of the power plant boiler as

$$ppm_v\,NO_x = a + bS \,, \qquad\qquad (24\text{-}3)$$

where S = size of boiler (MW/furnace firing wall) and the parameters a and b are as follows: gas fired, a = -17, b = 5.51, correlation coefficient r = 0.90, and standard deviation σ = 118 ppm_v; oil fired, a = 228, b = 1.59, and r = 0.59; coal fired, a = 291, b = 3.67, r = 0.95, and σ = 89 ppm_v.

Combustion modifications that have been tried include lowering the excess air, staging the combustion, recirculating some of the flue gas, fuel additives, and combinations of these. Fuel additives are not very likely to be used nowadays when concern is being expressed about all the fuel constituents except

hydrogen and carbon. Bartok et al. investigated the variations of
the other combustion parameters. They found that: Staging the
combustion would reduce the nitrogen oxides emitted by 22-54%,
averaging 31, 22, and 39%, respectively, for gas, oil, and coal
firing at 60-85% load; lowering the excess air resulted in reductions of 7-32%, averaging 19, 22, and 17% for gas, oil, and coal
firing; flue gas recirculation reduced emissions by 10-60% in
limited trials. A combination of techniques tested showed reductions of 51, 59, and 55% for gas, oil, and coal firing, respectively.
The controlling factor in combustion modifications is the point that
the carbon monoxide emissions start to rise beyond acceptable
limits.

B. Other Control Methods

Nitrogen oxides may be removed from the flue gas. The
removal methods, which are comparable to those for sulfur dioxide,
include alkaline scrubbing, activated charcoal adsorption and
catalysis, and the lead chamber acid process. The relatively
dilute concentrations of nitrogen oxides in the usual flue gases
and the low solubility of the nitrogen oxides present the same
processing problems as those described above for sulfur dioxide.

IV. AUTO EXHAUST

Motor vehicles are the sources of much air pollution; the
pollutants from vehicles include particulate matter, gases and
vapors, and noise. Controls have been applied for the noise from
engines and tires, for the particulate matter from the roadways,
and for some of the gaseous and vaporous emissions (crankcase,
carburetor, gas tank, and exhaust). The most vexing control
problems now being attacked are the further reductions of the
exhaust emissions, especially the oxides of nitrogen reduction.

A. Regulations

Regulations for auto exhaust emissions have been set at

levels which require the development of new control technology
before 1976 (see Table 24-2). The wording of the law for air
pollution sources other than auto exhaust is "the best available
technology that is feasible shall be used." Perhaps this approach
would have been more meaningful for auto exhaust as well. Certainly, the very stringent exhaust controls are needed in only a
few locations at this time; many of the management decisions for
auto exhaust control were based on oxides of nitrogen measurements
that were in error by a factor of two.

B. Meeting the Regulations

Although the intent of the early California regulations was
that a catalytic muffler would be used, auto makers met these
1969 standards by changing the engine operation. The hydrocarbon
and carbon monoxide emissions were lowered at the expense of increasing the nitrogen oxides and detuning the engine so that it is
hard to start, poor in performance, and high in gasoline usage.
The controls have been detuning, injecting air into the exhaust
manifold, and changing the air-to-fuel ratio (see Fig. 24-1). At
least one company is promoting a system to permit operating the
engine at a 19:1 fuel-to-air ratio, which would give low emissions
of all three of the major exhaust pollutants.

The 1963 model autos incorporated positive crankcase ventilation (PCV) to prevent crankcase blowby (~3.15 gm/mi), and
starting in 1971, carburetor and gas tank evaporations have been
captured in activated carbon canisters and returned to the carburetor (2.77 gm/mi reduced to <0.49 gm/mi).

Lead additives are being banned from gasoline. In 1974,
all gasoline stations must have lead-free gas available and within
a few years no leaded gasoline is to be allowed. The ban should
result in less lead in the environment and an exhaust gas stream

TABLE 24-2

Control/Regulation of Auto Emissions (gm/mi)[a,b]

Source/emission	Uncontrolled (before 1963)	Model year control/regulation effective							Thermal reactor	Catalytic converter
		1968	1971	1973	1974	1975	Interim[c]	1976		
Crankcase blowby (HC)	3.15	0[d]	--	--	--	--	--	--	--	--
Evaporation (HC)	2.77	--	0.49	--	--	--	--	--	--	--
Exhaust Hydrocarbons (HC)	17 (1500)[g]	4.1 (275)[g]	5	3.0	[1.5]^e	0.41	1.5[0.9]^f	0.41	0.45	0.31
Carbon monoxide (CO)	87 (33,000)[g]	34 (15000)[g]	45	28	[23]^e	3.4	15[9]^f	3.4	7.0	2.6
Nitrogen oxides (NO$_x$) (as nitrogen dioxide)	8 (2000)[g]	--	4	--	[2.0]^e	--	3.1	0.4	1.1	0.3

[a] "Rules and Regulations," Federal Register, 37:221, Part II, 24250-24320, November 15, 1972 and other sources.

[b] Regulations for trucks in 1974: 16 gm/BHP-hr for HC and NO$_x$ and 40 gm/BHP-hr for CO; i.e., for a truck with 100 BHP moving 50 mph, 32 gm/mi of HC and NO$_x$ and 80 gm/mi for CO.

[c] Standards to be substituted for the 1975 Federal statutory standards.

[d] Beginning in 1963.

[e] California state regulations.

[f] Number in brackets for California only.

[g] ppm$_v$.

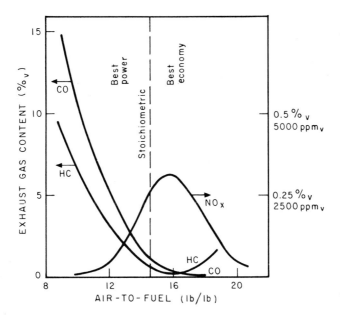

FIG. 24-1. Effect of fuel mixture on exhaust.

that is less poisonous to catalysts; there is no good evidence that the former is needed and there is increasing evidence that the latter may not be acceptable. The ban is very expensive in the refining process revisions and in the car modifications. Several catalyst combinations and configurations can meet the specified emissions levels, but achieving the specified 50,000 miles durability is proving difficult. Also, there is now concern over the sulfuric acid emissions from the catalytic mufflers and about the heavy metals released by these systems. Although there has been much experimentation with and publicity about the various metal oxides that can serve as catalysts, the plans now are to use the conventional noble metals catalysts.

Substitute power sources for vehicles are receiving major interest. These include the external combustion engine, the Sterling engine, the stratified combustion engine, the gas turbine,

and the electric propulsion by batteries or fuel cells. The only
one of these which is headed for significant early production is
the stratified charge engine developed by Honda through the re-
vision of a concept that has been around for many years; several
manufacturers have been licensed to produce this type of engine
for compact cars.

Whatever the ultimate solution to the automotive emission
control, the cost will be high. The cost of exhaust controls which
started out to be $50 and is now up to about $200 appears headed
for several hundred dollars per car in the near future.

V. METALLURGICAL FUMES

Molten metals give off metal vapors which react with atmospheric oxygen to form the metal oxides that condense to give small
particles (<0.1 µm)--fumes. There are currently three types of
air cleaners that can remove fumes at the efficiencies required by
modern regulations, namely, bag filters, electrostatic precipitators,
and high-energy scrubbers.

Because of the small size of the fumes, the cake that builds
on the bag filter has a high flow resistance. Filters applied to
fumes usually have air-to-cloth ratios (superficial velocities) on
the order of 1 fpm and sometimes even lower. This makes for
large and expensive installations; however, a properly operated
filter will clean the flue gas to the level of no visible emissions.
Such cleaners are the normal means of cleaning lead fumes. If
the new generation filters with pulsed jet cleaning can bring the
air-to-cloth ratios up to 3 or 4 fpm, filters will have an even more
commanding advantage in these applications.

Electrostatic precipitators work satisfactorily on some fumes,
especially iron and copper. They are not normally applied to lead
fumes. The prime requisite for good design is the use of realistic
migration velocities for the particles.

High-energy scrubbers have been used with increasing frequency during the last several years. The rise in popularity may have been caused to some extent by high interest rates that make initial low cost more attractive. Other factors involved in the selection of high-energy scrubbers have been the ease of increasing the efficiency of the cleaner after installation and their ability to handle the hot gases involved in metallurgy. Scrubbers on metallurgical flue gases characteristically operate at head losses in excess of 50 in. w.g., usually 60-80 in. w.g. and some have been operated at head losses near 100 in. w.g. The hot water/steam scrubbers described earlier are being applied to fume removal with reported high efficiencies. The principal difficulty in such application would seem to lie in boiler problems from recirculation of dirty (high sulfates) water or in disposal of this water for once-through use.

VI. ACID MISTS

Numerous demisters have been tried for collecting the fine mists (~0.3 µm) from acid manufacture. The highly obnoxious nature of such mists dictated high efficiencies of collection even before regulations were instituted. The first use of the electrostatic precipitator was on sulfuric acid mist. Electrostatic precipitators continue to be a good method for removal of acid mists. An alternative is the use of a deep-bed diffusion filter of fiberglass or a thin impaction-type demister. Many inertial collectors have been tried at low impactive power with predictable results for particles the size of acid mists.

VII. NOISE

At the same time that much concern has been given to sonic booms from supersonic aircraft, there has been an alarming increase in the outdoor noise levels. Gasoline engines on lawn mowers, chain saws, snowmobiles, and trail bikes have increased in number, power, and noise output. Ambient noise levels from industrial operations should be corrected by the institution of noise controls to meet the OSHA regulations.

Noise control may be obtained by modifying the equipment that produces the noise, by isolating the noise source, or by absorbing the acoustical power with insulation. Jet aircraft have been regulated for noise control purposes around airports. Quieter jets have been designed and flying patterns have been changed, sometimes to the detriment of safety. Hot-mix asphalt plants have produced high noise levels with their burners; now muffled air intakes can reduce this source of noise, but there is still a problem with noise emitted from the stack of the kiln. Noise from truck engines has not been well controlled. Efforts are underway to reduce train noises, both from the engines and the wheels. Highways are now being designed with some concern for the noise exposure to the residents nearby. Ambient noise regulations are needed in many areas and enforcement is needed in many areas where regulations now exist.

VIII. OTHER POLLUTION

The increasingly stringent air pollution regulations will extend required controls to more and more pollutants. These pollutants will generally be either gaseous or particulate in nature and will be controlled by conventional means; however, in many cases they will be physical agents such as noise, heat, and electromagnetic radiations. Some of the more unusual problems will be discussed briefly here.

A. Pollen

An estimated 10% of the population reacts allergically to pollen. Despite the problems associated with pollen, there has been little effort to control this pollutant. Full control of the sources of pollen is impossible; however, the plants that are purely nuisances could be destroyed to a large extent and replaced with nonallergenic plants. Until and unless efforts are made to destroy the sources of the pollen, the best control appears to be removal of the pollen from our immediate environment. The size of pollen particles (18-25 μm) makes them rather easy to remove by filtration, scrubbing, or electrostatic precipitation. Special home and office cleaners are being sold for the purpose, but mostly the removal is done by the central heating/air conditioning units. Automobiles have not generally been equipped with filters to remove pollen/dust from the intake air. Such removal presents no formidable problems and will probably be practiced in the future.

B. Cooling Tower Fog

Cooling towers have been used extensively in the past, and as a result of thermal water pollution regulations, their use is going to increase in the future. During ordinary operations, the only air pollution which might result from cooling towers would be the salt spray from towers using salt water or the hydrocarbons stripped from towers using refinery wastewater. Salt spray from cooling towers caused concern when the mist was estimated as high as 0.1% of the total circulated water. Measurements by Fish and others have shown the true value to be less than 1/30th of this amount. Use of polluted waste waters as cooling water may need to be discontinued.

During cold weather, cooling towers produce a steam fog that causes visibility and icing problems on adjacent highways. Some towers have been equipped with burners to alleviate these fog problems. Such burners are expensive to install and to operate (fuel for 1/3 to 1/2 of the Btu cooling capacity of the tower). Furthermore, they usually are too small to prevent the formation of the fog and work by lifting the fog over the adjacent highway. Cooling towers should be located as far as possible away from highways to allow dilution of the fog; towers may be replaced during brief periods of fog formation with pond cooling or once-through cooling; operations may be curtailed during fog-forming conditions.

C. Wastewater Treatment

Wastewater treatment practices continue to contribute to air pollution. Biological wastewater treatment injects large numbers of bacteria into the air; the air being emitted from aeration tanks contains more than 50 pneumonia-causing bacteria per cubic foot. These may present little hazard to the employees and other people nearby, but they need additional study. Hydrocarbon pollution has caused much concern, but the wastewater at refineries and petrochemical plants is given treatment that strips large quantities of hydrocarbons into the air. Some of the stripped material results from nonintentional actions such as the bubbling of air through the wastes for supporting biological degradation; however, if a waste is too concentrated for biological degradation, stripping is often done to lower the level to acceptable limits for the microorganisms. Ammonia stripping to the atmosphere is practiced at denitrification units for wastewater treatment. Along with the bacteria and hydrocarbons and ammonia, other odorous compounds are emitted from wastewater treatment, and the particles ejected into the atmosphere may be irritative to the respiratory tract as well as containing the bacteria.

D. Feedlot Emissions

Cattle production in this country employs putting large numbers of cattle on a small area and heavy feeding. The problems that arise from having 30,000 head of cattle on less than 40 acres of land are severe. The air pollution problems stem from the odors, noise, and dusts. People are becoming less tolerant of such conditions. A farmer in Texas recently won nearly $100,000 in damages for the reduction in value of his property which was close to a feedlot. Serious complaints extend to distances of more than a mile. Some control of the odors has been obtained by cleaning the lot, putting the cattle on concrete floors and washing the wastes into tanks, and oxidizing the odors or using perfumes. The air pollution problems from cattle feeding operations will probably continue until the cattle are completely enclosed and the air from the enclosure is cleaned before release.

IX. SUMMARY

Air pollutants generally fit into the categorical groupings of gaseous and particulate. The control method for a certain pollutant is normally one of the conventional control methods described in the earlier chapters of this book. Another group of air pollutants that must be considered are the physical agents, a group that includes noise, ionizing and nonionizing radiations, and climatic factors.

Control of sulfur dioxide and oxides of nitrogen from power plants will be difficult and expensive to achieve. These gases are present only in rather small concentrations and are contained in very large volumes of flue gas. Our technology for dealing with these gases has been that for the production of sulfuric acid and nitric acids from concentrations of several percent. The sulfur dioxide from smelting will be much easier to control than power plant gases because the concentrations are much higher. The oxides of nitrogen controls that prevent the formation of such gases stand a better chance of success than the removal of the oxides after they are formed. Coal gasification, then scrubbing the gas of the sulfur, appears to be much more feasible than the removal of the sulfur dioxide from the stack gas.

Auto exhaust controls that have been mandated for 1976 are quite rigorous. Catalytic mufflers may be used to control the exhaust from conventional internal combustion engines; however, there is a distinct likelihood that another type of engine will be substituted into cars of the future or that mass transit will become a reality.

Engineering controls for the particulate pollutants and for the usual gaseous pollutants will not be too difficult. It will be expensive, and the degree of expense will depend to a large extent on the degree of cleaning that is demanded. Removal of the last vestige of a pollutant is most expensive.

PROBLEMS

1. Prepare a brief literature survey of the activated alumina process that the Bureau of Mines has tried to develop for sulfur dioxide removal.

2. Compare the nitrogen oxide emissions calculated by Eq. 24-3 with the empirical relation that

   NO (lb/10^6 Btu fired) = $(°F/5340)^{0.535}$ (% excess air + 38).
   Use a 400-MW thermal, horizontal opposed firing boiling with pulverized coal (200 MW/furnace firing wall), 2800°F firebox temperature, and 30% excess air.

3. Explain why the cost of auto exhaust cleanup has been estimated at \$27/ton for the first 67%, \$94/ton for the next 23%, and \$740/ton for the next 9%.

4. What is the probable efficiency of removal of 0.3-μm sulfuric acid particles by an inertial scrubber which has 7.5 HP for a 10,000-cfm unit (From current advertising.)?

5. What are the possible noise control methods for use on a 2000-HP fan for a venturi scrubber?

BIBLIOGRAPHY

Bartok, W, A. R. Crawford, E. H. Manny, and G. J. Piegari, "Reduction of Nitrogen Oxide Emissions from Electric Utility Boilers by Modified Combustion Operation," presented at American Flame Days, Chicago, September 1972.

Beranek, L. L., Ed., Noise and Vibration Control, McGraw-Hill, New York, 1971.

Chopey, N. P., "Coal Gasification: Can It Stage A Comeback?" Chem. Engr., 79:7, 44-46, April 3, 1972.

Control of Sulfur Oxide Emissions in Copper, Lead, and Zinc
    Smelting, Bureau of Mines IC 8527, Washington, D. C.,
    1971.

Control Techniques for Carbon Monoxide, Nitrogen Oxide, and
    Hydrocarbon Emissions from Mobile Sources, National Air
    Pollution Control Administration Publ. No. AP-66, Wash-
    ington, D. C., March 1970.

Davis, J. C., "Scrubber-Design Spinoffs from Power-Plant Units?"
    Chem. Engr., 79:28, 48-50, December 11, 1972. "SO₂
    Absorbed from Tail Gas with Sodium Sulfite," Chem. Engr.,
    78:27, 43-45, November 29, 1971. "SO2 Removal Still
    Prototype," Chem. Engr., 79:13, 52-56, June 12, 1972.

Dennis, C. S., "Potential Solutions to Utilities' SO2 Problems
    in the 70's," Combustion, 42:4, 12-21, October 1970.

Hirai, M., R. Odello, and H. Shimamura, "Solvent/Catalyst
    Mixture Desulfurizes Claus Tailgas," Chem. Engr., 79:8,
    78-79, April 17, 1972.

Horlacher, W. R., R. E. Barnard, R. K. Teague, and P. L. Hayden,
    "Four SO2 Removal Systems," Chem. Engr. Prog., 68:8,
    43-50, August 1972.

Industrial Noise Manual, 2nd ed., American Industrial Hygiene
    Assoc., Detroit, Michigan, 1966.

Mackinnon, D. J., and T. R. Ingraham, "Minimizing NOₓ Pollut-
    ants from Steam Boilers," J. Air Pollution Control Assoc.,
    22:6, 471-472, June 1972.

Randall, C. W., and J. O. Ledbetter, "Bacterial Air Pollution from
    Activated Sludge Units," Amer. Industrial Hyg. Assoc. J.,
    27:6, 506-519, November-December 1966.

Ross, R. D., Air Pollution and Industry, Van Nostrand Reinhold,
    New York, 1972.

Shah, I. S., "MgO Absorbs Stackgas SO2," Chem. Engr., 79:14,
    80-81, June 26, 1972.

"SO2 Removal Technology Enters Growth Phase," Environmental
    Sci. Tech., 6:8, 687-691, August 1972.

"Synthetic Fuels: What, When?" Chem. Engr., 79:8, 62-64,
    April 17, 1972.

Strauss, W., Industrial Gas Cleaning, Pergamon Press, New York,
    1966.

Veldhuizen, H., and J. O. Ledbetter, Cooling Tower Fog: Control
    and Abatement, Report to Sinclair Refining Co., EHE 69-14:
    AP-2, The University of Texas at Austin, August 1969.
    "Cooling Tower Fog: Control and Abatement," J. Air Pollution
    Control Assoc., 21:1, 21-24, January 1971.

Appendix A

PROBLEM SOLUTIONS FOR PART A

CHAPTER 1

1. $A \times P = \pi[8006(5280)]^2[14.7(144)]/2000 = 5.94(10^{15})$ tons

2. @ $20^{\circ}C$, saturation $= 2.31\%_v$ Table 1-1 values by 0.9769
   N_2, 76.28; O_2, 20.46; H_2O, 2.31; A, 0.91; CO_2, 0.03

3. $\dfrac{0.25 \times 454,000}{[1000/(0.075 \times 35.3)]} = 300$ mg/Nm^3

4. $0.25(7000)/(1000/0.075) = 0.13$ gr/Ncf

5. From Table 1-5, $1.7(10^6)$ tons/yr
   $1.7(10^6)(2000)(210 \times 10^6) = 7.14(10^{17})$ yr^{-1}
   Assume: avg. lifetime = 4 hr; areal extent = $5(10^6)$ mi^2
   vert. extent = 1000 ft (for avg. concentration); seasonal
   peak = 5 X avg.

   $\dfrac{7.14(10^{17})(4/8760)(5)}{5(10^6)(5280)^2(1000} = 0.012$ ft$^{-3} = 0.41$ m^{-3}

   [Note: Actually the number of pollen grains per pound is
   probably 40 times the reference value; ~8000(10^6) lb^{-1}.]

CHAPTER 2

1. No. Because 15-50%$_w$ of the kiln dust is smaller than 3 μm
   and much of this fraction would escape collection.

2. The usual pollution emissions listings show about 250(10^6)
   tons/yr for the U. S. (see Table 1-5) while the CO_2 emissions for the U. S. are near 10,000(10^6) tons/yr.

3. A unit of power, such as the watt. For total noise generation,
   a large multiple of the watt, such as the gigawatt or tera-
   watt, might be more manageable.

4. Odors, thermal pollution, and electromagnetic radiation are
   examples.

5. Radioactivity, benzopyrenes, PAN, heavy metals such as
   mercury, lead, vanadium, etc.

6. Boats and ships, construction, sludge drying, recovery units.

251

CHAPTER 3

1. $u = 12(2/10)^{1/7} = 9.54$ knots

2. 0–100 ft, inversion (very stable); 100–400 ft, unstable; 400–1000 ft, stable; 1000–1500 ft, isothermal (stable); 1500–2500 ft, stable; 2500–6000 ft, neutral.

3. $77^O F - 5.4 H$ = temperature at H, where H in 1000's ft. Dry adiabat from $77^O F$ will not intersect temperature structure; therefore, maximum mixing depth is greater than 6000 ft, and it may be unlimited.

4. $(2 \tan 5^O)(5280 \times 10)(1200) = 1.11(10^7)$ ft^2 (plume x-section)
   Dilution volume = $1.11(10^7)(10 \times 22/15)(60) = 9.76(10^9)$ cfm

   Dilution = $4 \times 10^5/(9.76 \times 10^9) = 4.10(10^{-5})$
   This value falls within the band of the figure. It is near the top of the band or the minimum expected dilution.

5. $\sigma_y = 350$ m and $\sigma_z = 50$ m (from Fig. 3-12)
   $Q = 400,000$ cfm = 189 m^3/sec; u = 10 mph = 4.47 m/sec
   $X = 189 \times 10^6/[\pi(350)(50)(4.47)]$ exp$[-1/2 \ (60^2/50^2)]$
   $= 374$ ppm$_v$ A dilution of 2673 times or $3.74(10^{-4})$.

6. Add 2.4% to <3 mph for SE.
   4.3%, <3; 10.4%, 4–7; 7.7%, 8–12; 1.9%, 13–18; 0, >18.
   $H_e = 100$ m = 328 ft; $Q_{sd} = 300$ tons/day = 41.8 ft^3/sec
   $x_m = (H_e^2/C_z^2)^{1/(2-n)}$; $X_m = 2 \ Q(10^6)C_z/(\pi e u H_e^2 C_y)$
                                        $= 91.02/u$ ppm$_v$

   Above 25 m, $C_y = C_z = C$

u (mph/fps)	C (ft$^{n/2}$)	n	x_m (ft)	X_m (ppm$_v$)
2/2.93	0.045	0.40	67,350	31.0
5.5/8.07	0.06	0.30	25,000	11.3
10/14.7	0.08	0.25	13,500	6.2
15.5/22.7	0.14	0.20	5,550	4.0

   Dose curve: Dist. (ft)/Dose (ppm$_v$-hr)--5000/900; 10,000/4000; 20,000/12,400; 50,000/17,000; 100,000/12,700; 200,000/5600; 400,000/1750.

   [Note: A slip on the intended 400-m H_e value made this problem very academic--these are indeed very high concentrations and doses! Also, the dispersion equations should not be used at large distances downwind.]

CHAPTER 4

1. Geometric mean. One count differs widely from the other two;

in order to prevent undue influence of the large value, the geometric mean should be used. ($\bar{x}$ = 43.3 vs. $\bar{x_g}$ = 36.3)

2. Σf = 100; Σfm = 7195; Σfd^2 = 2625; Σfd^3 = 1019; Σfd^4 = 178,290

$\bar{x}$ = 7195/100 = 71.95 $\mu gm/m^3$

σ = $[2625(100/99)]^{1/2}$ = 5.15 $\mu gm/m^3$

Sk_p = $(1019/100)/(2625/100)^{3/2}$ = 0.076

Kurtosis = $(178,290/100)/(2625/100)^2$ = 2.59

3. Size/%<: 64/7; 69/30.7; 74/70.3; 79/91.1; 84/99

$\bar{x}$ = x_{50} = 71.5 $\mu gm/m^3$ σ = x_{84} - x_{50} = 5.3 $\mu gm/m^3$

4. $\bar{d_g}$ = x_{50} = 28.5 um σ_g = 28.5/12.4 = 2.30

90% conf. limits = 28.5 $\mu m \stackrel{x}{\div} 2.30^{1.64}$ = 28.5 $\mu m \stackrel{x}{\div} 3.92$

5. N = 6 Σxy = 931.8 Σx = 63 Σy = 109.2 Σx^2 = 819

$r = \dfrac{(6 \times 931.8) - (63 \times 109.2)}{(6 \times 819 - 63^2)^{1/2}(6 \times 2281 - 109.2^2)^{1/2}}$ = -0.9983

6. 34 $\pm$ $(0.4^2 + 0.1^2)^{1/2}$ = 34 $\pm$ 0.41 ml

CHAPTER 5

1. Gas must stay in gaseous state. Equation works for most practical problems.

2. Filter first, then critical orifice, then pump if the pump is capable of producing a vacuum of more than 28.75 in. Hg; otherwise, the orifice on the pressure outlet of the pump and the filter on the vacuum side of the pump.
The area of the filter openings is unknown; however, in many practical applications, flow through a given type of filter remains almost constant for a given head.

3. The fiber glass immobilizes the air to prevent convective heat transfer.

4. Retentivity of carbon for phenol (x/M) = 30%$_w$

x/M = $kc^{1/n}$ Assume n = 2 and c = 10 ppm_v.

Vapor pressure @ 0°C = 0.12 mm Hg = 157.9 ppm_v

$0.30 = k(157.9)^{1/2}$ k = 0.0239 $ppm_v^{-1/2}$

$x/M = 0.0239(10)^{1/2}$ = 0.0756   x > 6 μgm M > 80 μgm

No. Molecular weight of ethylene is 28. Not readily condensable. b.p. = -103.9°C

5. Intermediate regime. u_t = 0.34$(5.24)^{2/3}$(10) = 10.26 cm/sec

ω = 600(10.26) = 6155 cm/sec

$N_{Re} = (1.2 \times 10^{-3})(6155)(10 \times 10^{-4})/(1.8 \times 10^{-4}) = 41$

Motion is well within the intermediate range.

6. $\Omega = [0.884(10^{-4})(16/2.54)^2(1)/(1.8 \times 10^{-4})][6/(6+2)] C_c$

   $= 16.35$ cm/sec $C_c = 1.12$ $\xi \simeq 6$ for limestone

7. $v = 3960$ ft/min $= 2012$ cm/sec

   $N_{Re} = 0.0012(45.72)(2012)/(1.8 \times 10^{-4}) = 6.13(10^5)$

   $\Delta P/L = 0.032(48.0)(528)(8.75)^2/[40(2117)(1.5^2)]$

   $= 0.10$ lb/ft^2-ft

   $\Delta P = 0.10(70)(12)/62.4 = 1.30$ in. w.g.

CHAPTER 6

1. $C/C_0 = (t/t_0)^\alpha$ $\alpha \simeq -0.35$ $C_{0.5} = 604(0.5/24)^{-0.35}$

   $= 2340$ μgm/m^3

2. $(0.5 \times 10^{-6}$ m^3/m$^3)(94 \times 10^6$ μgm/0.024 m$^3) = 1958$ μgm/m^3

   Several micrograms could be obtained in 0.005 m^3.

3. Adsorb on activated carbon or condense. Impinger or tube packed with activated carbon. At least 2 min @ 2.8 lpm.

4. Hi-vol (8 in. X 10 in.) fiber glass filter for sampling. Soxhlet extractor and colorimeter for analysis.

   $1.64(0.25)(10.6)/\!/N \le 0.25(10.6)$ $N \ge 2.69$ m^3

   Average city concentration of 10.6 μg/m^3 from Table 1-4.

5. $h = (10 \times 3.28)^2/(2 \times 32.2) = 16.7$ ft of air $\rho_{air} = 0.0489$ lb/ft^3

   $= 16.7(0.0489 \times 12)/(0.6 \times 62.4) = 0.26$ in. red oil

6. Velocity of sound $= 1087(293/273)^{1/2} = 1126$ ft/sec (Table 1-2)

   Area $= 0.1$ cfm/(1126 X 60 fpm) $= 1.48(10^{-6})$ ft^2

   $D = 0.00137$ ft $= 0.0165$ in. $= 0.418$ mm

   Midget impinger operates @ $(0.418/1)^2 = 0.175$ or 18% sonic.

7. $740 = 2000 \exp(-0.4k)$ $k = 2.49$ hr^{-1}

   $kt = 1$ for $C/C_0 = 0.37$; 20 ft^3 in 24 min or 0.833 cfm

8. $CE = 720$ $C(1 - E) = 24$ $E = 0.968$ Assumes $E_1 = E_2$.

9. 3-ft direction: $36/4 = 9$ in.
   4.5, 13.5, 22.5, 31.5 in. from side
   4-ft direction: $48/4 = 12$ in.
   6, 18, 30, 42 in. from side

CHAPTER 7

1. $C = C_0 + p\,(dC_0/dt) = 0.75 + 3.5\,(0.2) = 1.45$ ppm$_v$

2.
City	$\overline{x}_g/\sigma_g$	Population
Chicago	0.09/3.33	$3.5\,(10^6)$
Philadelphia	0.04/3.75	$2.0\,(10^6)$
St. Louis	0.04/2.75	$0.7\,(10^6)$
Washington	0.035/2.57	$0.76\,(10^6)$
Cincinnati	0.02/2.75	$0.48\,(10^6)$
San Francisco	0.008/2.88	$0.73\,(10^6)$

3. $I/I_0 = \exp(-kcL)$ $L = 1$ cm $k = 2170/56\,(10^6)$
                                                        $= 3.875\,(10^{-5})\;\ell/\mu g$
   % T/c   100/0 $11.4/56\,(10^3)$

4. Fe^{+++} in mole/ℓ: 0 0.001   Scales: 1 Absorbance unit
   Absorbance (o.d.): 0 1.06   for 0.001 mole

5. It gives a balance of oxidation/reduction and permits measurement of the basic potential for the Nernst equation.

6. 3 cycle semi-log
   PIT$_{50}$ (PRT$_{50}$): $C = 1$ ppm$_v$ @ Perceived Intensity = 1
                          $C = 100$ ppm$_v$ @ Perceived Intensity = 4
   PPT$_{50}$: Perceived Intensity = 1/2 @ $C = 0.465$ ppm$_v$

CHAPTER 8

1. Along the line of sight because particles will lie flat most often.

2. $\overline{d} = \Sigma fd/\Sigma f = 1066.5/106 = 10.06\;\mu m$

3. $\ell n\,MMD = \ell n\,CMD + 3\,\ell n^2\,\sigma_g = 0 + 3\,(1.099)^2 = 3.621$
   $MMD = anti\ell n\,3.621 = 37.4\;\mu m$

4. $\ell n\,CMD = \ell n\,MMD - 3\,\ell n^2\,2 = \ell n\,MMD - 1.4414$
   MMD/CMD: 10/4.50; 5.8/1.37; 1.7/0.40

5. The beaker for the 2-μm particles may be decanted only $(2/5)^2$ as deep as the 5-μm particle beaker; the beakers may be decanted to the same depth by using beakers with the area ratios 4:25.

6. $u_t = 0.003\rho_p d^2$ p. 182 $u_t = 0.003\,(2.65)\,(10)^2 = 0.80$ cm/sec

7. $25,000 = (24,000\pi D/60)^2/(32.2\,D/2)$ $D = 0.255$ ft = 3.06 in.

8. Resolution is an inverse function of wavelength (λ) and the wavelength for UV is less than for visible light.

9. $\sigma = \sqrt{480} = 21.9$ $[480 \pm 1.64\,(21.9)]\,200 = 96,000 \pm 7180$ ml^{-1}

10. $\overline{d^2} = \Sigma d^2/N = 31,757/106 = 299.6\;\mu m^2$

11. Assume $A = 10 \text{ cm}^2 = 10^9 \, \mu m^2$ and $\overline{d^2} = 300 \, \mu m^2$.

$N_0 = 0.2(10^9)/(300\pi) = 212,500$ [Actual count = 212.5/field]

CHAPTER 9

1. $X = 2 \ 7/16$ in. $= 3.85(10^{-5})$ mi $\sigma_X = 1/32$ in. $= 4.93(10^{-7})$ mi
    $Y = 0.048$ gm $= 5.29(10^{-8})$ ton $\sigma_Y = 0.003$ gm $= 3.31(10^{-9})$ ton
    $Q = (Y/1.5)/(\pi X^2/4) = 0.8488 \, Y/X^2$
    $\sigma_Q{}^2 = (\partial Q/\partial X)^2 \sigma_X{}^2 + (\partial Q/\partial Y)^2 \sigma_Y{}^2$
    $\quad\quad = [0.8488(-2)Y/X^3]^2 \sigma_X{}^2 + [0.8488(1)/X^2]^2 \sigma_Y{}^2$
    $\quad\quad = 0.602 + 3.593 = 4.195 \quad\quad \sigma_Q = 2.05$
    Dustfall $= 30.3 \pm 2.05 \text{ ton/mi}^2\text{-mo}$

2. $X(1.6)(200) = 30(1000) \quad\quad X = 93.8$ min

3. 55% water vapor @ 550°F   50 ppm$_v$ SO_3 (see Fig. 9-1)
    Abscissa $= 50/550,000 = 0.91(10^{-4}) \quad \Delta t = 50°C$
    Theoretical dewpoint: $\log(0.55 \times 760) = 8.0573 - 1724/(T - 40.13)$
    $\quad\quad\quad\quad\quad\quad\quad\quad\quad\quad\quad T = 357°K = 84°C$
    Actual dewpoint $= 84° - 50° = 34°C$

4. $2(60)(10/28.32) = 42.4 \text{ ft}^3 \quad$ Area $= \pi (9/16)^2/144(4) = 0.00173 \text{ ft}^2$
    $L = \text{volume/area} = 24,550 \text{ ft} \quad$ Absorbance $= 0.7$
    S.I. $= 10^5 \, A/L = 10^5 (0.7)/24,550 = 2.85$ Cohs

5. See reference <u>Selected Methods for the Measurement of Air
    Pollutants</u>

    It has been empirically determined that 1 mole of NO_2 is
    equivalent to 0.72 mole of $NaNO2$ in producing color.
    $[(2.03/100)/(0.72 \times 69)](24.5 \times 10^6)(0.001) = 10.01 \, \mu l \, NO_2/ml$
    The mole volume used is for 25°C; for 20°C use 2.06 gm/l.

6. See reference above. 1 mole of ozone liberates 1 mole of iodine.
    $0.0025 \text{ N } I_2 = 0.00125 \text{ M } I_2 \quad$ 1 ml of I_2 solution $= 1.25 \, \mu$mole
    $1.25(24.5 \times 10^6) = 30.6 \, \mu l$ ozone/ml of std. @ 25°C
    Standards are for 25 ml; sample is 10 ml; $30.6(10/25) = 12.24$

CHAPTER 10

1. N_2, 77.19%$_v$; O_2, 20.71%$_v$; A, 0.92%$_v$; CO_2, 0.03%$_v$; H_2O, 1.15%$_v$.

2. $Ct = K$ 400(2) = 800 ppm_v-hr   t for 1000 = 0.8 hr = 48 min

3. 1962 $185(10^6)(10/10^5)$ = 18,500 deaths   3 X 1952 number.

4. 5 ppm_v SO_2 lengthened life of guinea pigs and lowered incidence and severity of lung lesions. APCA J., 645, 1969 (September).

   3000 R x ray exposure lengthened life of flour beetles 10%.
   0.8 R/day of Co-60 gamma rays lengthened life of rats 30%.
   0.01 μCi/gm Sr-90 injection reduced bone tumors in mice.
   Bacq and Alexander, Fundamentals of Radiobiology, 2nd ed., pp. 439-443, Pergamon Press, London, 1961.

   0.1 R/day of Co-60 gamma rays lengthened life of mice 6-7%.
   Andrews, Radiation Biophysics, p. 278.

5. $1.6(10^6)(0.417)$ = 667,000

6. $R_v = v = 2.996(2 \rho d)/(3 K_s W)$ k = vW = 5.992(2.5)(0.682)/(3 X 2)

   $$= 1.70 \text{ gm/m}^2$$

   [Note: Units correct because ρ to gm/m^3 takes multiple of 10^6 and d to m takes 10^{-6}.]

7. $R_v = k/W$ $k = 1.8 \text{ gm/m}^2$ R_v = 5 mi = 8047 m   w = 224 $\mu g/m^3$

8. Number of particles = $(1/10^{-5})^3 = 10^{15}$

   Area per particle = $6(0.1)^2 = 0.06 \mu m^2$

   Total area = $0.06(10^{15})/10^8 \text{ cm}^2 = 60 \text{ m}^2$

9. R_v (mi) = 11 exp(-0.479 X 6) = 0.621 mi = 1 km

   W = 1.8/1000 = 1800 $\mu g/m^3$

10. Concentration = buildup - decay = $C[1 - \exp(-0.693 \, t/T_{\frac{1}{2}})]$

    = 5(.03/12)(1 - exp -3.46) = 0.012 ppm_v

11. $[COHb]/[O_2Hb] = R \, PP_{CO}/PP_{O2}$

    $$\% \text{ COHb} = \frac{100 \text{ COHb}}{COHb + O_2Hb} = \frac{100 \, R \, PP_{CO}/PP_{O2} \, [O_2Hb]}{R \, PP_{CO}/PP_{O_2} \, [O_2Hb] + [O_2Hb]}$$

12. $40 = 100 \, PP_{CO}/(210,000/300 + PP_{CO})$ PP_{CO} = 467 ppm_v

13. Cotton is very susceptible to ethylene damage; cotton growing adjacent to plant is good public relations evidence of low emissions of ethylene.

14. $\rho_p = 1.2 \text{ gm/cm}^3$ vol./part. = $\pi d^3/6 = 6.54(10^{-8}) \text{ cm}^3$

    $(1)(454)/[1.2 \times 6.54(10^{-8})] = 5.78(10^9)$ particles/lb

15. 600 rads X 100 ergs/gm-rad X $2.39(10^{-8})$ cal/erg X 1 °C/cal-gm

    = 0.0014°C

Appendix B

PROBLEM SOLUTIONS FOR PART B

CHAPTER 11

1. Alfalfa damage by SO_2: The bulk of the plant, especially the leaves, is the valuable part of the alfalfa; therefore, the criteria should deal with the amount of weight reduction by SO_2 exposure (leaf drop)--how much SO_2 for how long to cause a given damage.

   Gladioli damage by ozone: The appearance of the gladioli flowers and leaves makes the plants valuable; criteria should describe damage to appearance, also number of flowers.

   Citrus fruit damage by photochemical smog: Criteria should primarily be concerned with the reduction in the amount and quality of fruit produced. Damage to trees early in the season might affect the amount of fruit put on by the trees and late in the season might affect the amount of fruit drop (loss); exposure at either time could damage the appearance of the fruit.

2. See Problem 10-4. Add the following:
   6 R/wk of Co-60 gamma rays reduced the mortality among guinea pigs. Fundamentals of Nuclear Energy Research--1963, U. S. Atomic Energy Commission, p. 44.
   Rate of growth enhanced by gamma rays. Gloyna and Ledbetter, Principles of Radiological Health, Marcel Dekker, 1969. p. 187

3. Criteria for ozone: Humans--0.2 ppm_v for 3 hr/day, 6 day/wk over 12 wk caused no apparent effects; 0.5 ppm_v did. Odor, PRT_{90} = 0.02 ppm_v. Increased airway resistance, 0.6 ppm_v for 2 hr. Animals--0.08 ppm_v for 3 hr increased susceptibility to bacterial infection. Plants--0.05 ppm_v for 4 hr, leaf injury to sensitive plants.

   Guide for ozone: AIHA J., 29:3, 299-303, May-June 1968. Humans--Drying of nose, mouth, and throat at >0.1 ppm_v and general irritation at 1 ppm_v. Pulmonary edema from 1.5-2 ppm_v for 1 hr or more. Generally disagreeable odor at 0.05-0.5 ppm_v. Animals--0.15 ppm_v hastens death with streptococcal infections. Plants--0.053 ppm_v tobacco damage with 2-hr exposure.

These two sources are in better agreement for ozone than for any other pollutant, primarily as a result of similar approaches in their literature surveys. The recommended standard by the guide is for 0.1 ppm$_V$ to occur not more than 1 hr daily on an average for a year. The primary Federal standards set by the criteria list 0.08 ppm$_V$ maximum 1-hr concentration not to be exceeded more than once in a year. Some measurements of natural ozone show concentrations well above 0.1 ppm$_V$ from an apparent down-welling of the ozone layer.

4. Will vary from place to place, but must be equal to or less than those shown in the Federal ambient air quality standards (see Table 11-1).

5. Geometric standard deviation $(\sigma_g) = 1.74$ from Table 1-4 data.

   $C/C_0 = (t/t_0)^\alpha$ $\alpha \simeq -0.2$ (Chap. 6)

   $C_m = 60(1/365)^{-0.2} = 195$ ug/m^3

   $\overline{x_g} \div \sigma_g^z$ $z = 2.78$ for $1/365 = 0.0027$ (Probability tables)

   $z = 2.33$ for $1\% = 0.01$

   $195(1.74^{2.33}/1.74^{2.78}) = 152$ ug/m^3

   The state of Texas has a regulation using the 60 and 125 values that employs a different α from the one above. The Federal secondary standards show 60 and 150, which is in agreement.

   References: R. I. Larsen, APCA J., 23:11, 933-940, November 1973. EPA Publ. No. AP-89.

6. TLV $= 10$ ppm$_V$ $1/30(10) = 330$ ppb$_V$ PRT$_{50} = 5$ ppb$_V$ (Table 7-3)

CHAPTER 12

1. $\dfrac{0.012 \text{ gr}}{acf[(460 + 68)/(460 + 950)]} = 0.032$ gr/Ncf

   $[0.032(35.3)/7000]454,000 = 73.3$ mg/Nm^3

2. $0.032/(1.00 - 0.17) = 0.0386$ gr/Ncf (dry)

   $73.3/0.83 = 88.3$ mg/Nm^3 (dry)

3. $C/C_0 = (S/S_0)^{0.6}$ $C = \$150,000(70,000/150,000)^{0.6} = \$94,950$

4. $q = 1.00 - 0.80 = \exp(-k5^2)$ $k = 0.0644$ for $d < 10$ um

   $q = \exp(-kd)$ for $d \geq 10$ um $k = 0.644$ um^{-1}

   $3.7\% < 2$ um, use 1.2 um effective diameter

   $20\% > 10$ um by extrapolation on log X probability plot, use 12 um effective diameter

   $q = 0.20\exp(-0.644 \times 12) + 0.23\exp(-0.064 \times 7.5^2)$

   $+ 0.32\exp(-0.0644 \times 4.5^2) + 0.15\exp(-0.0644 \times 3.5^2)$

$+ 0.063 \exp(-0.0644 \times 2.5^2) + 0.037 \exp(-0.0644 \times 1.2^2)$

$q = 0.237 \qquad \eta = 0.763$ or $76\%_w$

5. $\eta = 81\%_w$

6. $q = \dfrac{1}{\sqrt{2\pi}\,\ell n\,3} \displaystyle\int_{-\infty}^{t} e^{-t^2/2}\, e^{-0.0644(4^2)e^{2\,t\,\ell n\,3}}\, dt$

for MMD = 4 μm and $\sigma_g = 3$

$t = (\ell n\,x - \ell n\,\overline{x_g})/\ell n\,\sigma_g = 0.83$ for $x = 10$ um

a	b	c	d	e	f	g
t	$\dfrac{1}{\sqrt{2\pi}}\displaystyle\int_{-\infty}^{t} e^{-t^2/2}$	$e^{2.2t}$	$\dfrac{b}{1.1}e^{-1.02c}$	$d_{avg.}$	Δt	$e \times f$
-4.0	0.00003	0.0002	0.0000			
-2.2	0.0139	0.0079	0.0125	0.0063	1.80	0.0113
-1.82	0.0344	0.0182	0.0307	0.0216	0.38	0.0082
-1.26	0.1038	0.0625	0.0885	0.0596	0.56	0.0334
-0.50	0.3085	0.6285	0.1477	0.1181	0.76	0.0898
0.00	0.5000	1.0000	0.1639	0.1558	0.50	0.0779
0.83	0.7967	3.6620	0.0173	0.0906	0.83	0.0752
		$e^{1.1t}$	$\dfrac{b}{1.1}e^{-2.56c}$			
0.83	0.7967	1.9136	0.0059			
1.26	0.8962	3.9888	0.0000	0.0030	0.43	0.0013

$q = \Sigma g = \quad 0.2971$

$\eta = 0.70$ or $70\%_w$

CHAPTER 13

1. $V = 6000[S/(S + 1)]d^{0.4} = 6000(5.2/6.2)(400/2540)^{0.4} = 956$ fpm

2. $Q = LVX = 2.8(4)(100)(7) = 7840$ cfm

3. $0.58 = 0.058(100)(6000^{1.91})/(C\,21^5) \qquad C = 40.3$

4. $h = aQ^b/D^c$ Use: $D = 24, 20,$ and 12 in.; $Q = 10,000, 3000,$
   and 5000 cfm. $h = 0.5, 0.12,$ and 4.35 in. w.g.
   $h = 0.077Q^{1.95}/D^{5.07}$ Eq. 13-8: $h = 0.1055\,Q^{1.91}/D^5$
   Appears to fairly good agreement.

5. Try $D = 5.5$ in. for Branch A: VP = 1.24 in. w.g. $h_e = 2.31$ VP
   Equiv. length of a and b: $L = 0.87(5.5)^{1.183} = 6.54$ ft
   Total equiv. length = $10 + 15 + 2(6.54) = 38.1$ ft
   Total head = $1.00 + 2.31 + 38.1/[0.0365(55)(5.5)(800)^{0.09}]$
   $= 5.20$ VP = 6.45 in. w.g. Balances with 6.48 in. w.g.

Branch C: Q = 800 + 600 = 1400 cfm

Minimum D: $1400/(\pi D^2/4) = 4000$ D = 0.67 ft = 8 in.

VP = 1.00 in. w.g.

Total head: $8/[0.0365(55)(8)(1400^{0.09})] = 0.26$ VP

$= 0.26$ in. w.g.

Branch D: D = 8 in.

Total equiv. length = $8 + 3 + 0.87(8)^{1.183} = 21.2$ ft

Total head: $21.2/[0.0365(55)(8)(1400^{0.09}) = 0.69$ VP

$+ 0.60$ VP for exit $= 1.29$ VP = 1.29 in. w.g.

Total head for system:

$6.48 + 0.26 + 1.29 + 8.00 = 16.0$ in. w.g.

Fan HP: HP = Qh/(6356 E) = 1400(16)/(6356 X 0.60) = 5.87

6. $0.645(14,700)(528/1260)(0.075)(0.24)(800^\circ - 275^\circ F) = 37,547$

$0.355(14,700)(528/1260)(0.046)(0.48)(800^\circ - 275^\circ F) = 25,570$

Total Btu/min   = 63,117

Area required:

$$\frac{63,120(60)}{2.15\{[(800-100)-(275-100)]/\ln[(800-100)/(275-100)]\}} = 4650 \text{ ft}^2$$

CHAPTER 14

1. Hooding individual pots collects the pollutants in a much
    smaller volume and does so with less exposure to the people
    working in the building.
    Using the building as a hood permits easier access to the pots.

2. The same as for Problem 1 plus the building for the open hearths
    is usually much more open; therefore, the volume would be
    quite large for the open hearths. A recent installation on
    electric furnaces uses very large, low-velocity hoods far
    above the furnace; the fumes rise with the heat of the furnace
    and the volume of air is very reasonable.
    The open hearths require fewer workmen and less hand labor in
    the building than the Soderberg pots.

3. Smelters, portland cement plants, acid plants, synthetic rubber
    plants, refineries, and petrochemical plants. Any industries
    that are relatively bad for pollution emissions and require a
    not too large work force.

4. Apparently the principal reason is that economics has not dic-
    tated chemical separation. A process by Treadmill Corp. is
    supposed to work as follows:

90% sulfuric acid leaches sulfide ore to form copper sulfate plus other sulfates and elemental sulfur; copper sulfate plus hydrogen cyanide gives copper cyanate; copper cyanate plus hydrogen gives 99.99% pure copper plus hydrogen cyanide.

Pure copper cannot be obtained by electrolysis of the ore put into solution without other treatment.

Advantages claimed for the chemical separation: Recovers 99.6% of the copper versus 97% for smelting; Converts 90.5% of sulfide to elemental sulfur; Uses nearly all of the sulfur dioxide generated; and Costs about 60% as much for capital costs and about 40% as much for operating costs as smelting.

5. Both produce noise, dust, and smoke. The noise from ammonium nitrate blasting should be somewhat less because such large quantities of the material are required that overloading of the holes is less frequent. The smoke from TNT is much worse than that from the ammonium nitrate; it produces a severe headache upon exposure. The dust put into the air will not be much different for the two processes.

CHAPTER 15

1. $Q = 20,000$ lb/hr $= 0.947$ m^3/sec

From Fig. 15-3: $x_m \approx 28$ km $X_m u/Q \approx 2.5(10^{-7})$ $X_m \approx 0.062$

$$X = \frac{Q\,10^6}{\pi \sigma_y \sigma_z u} \exp(-400^2/2\sigma_z^{\,2}) = 79,300/(\sigma_y \sigma_z)\exp(-80,000/\sigma_z^{\,2})$$

Distance (km):	5	10	20	28	50	100
σ_y (m):	300	550	1000	1340	2200	4000
σ_z (m):	88	136	205	243	330	460
X (ppm$_v$):	0.0001	0.014	0.058	0.063	0.052	0.030

2. $x = 1.5$ km, $\sigma_y = 100$ m; $x = 100$ km, $\sigma_y = 4200$ m

$\sigma_y = 69.7 x^{0.89}$

$x = 1$ km, $\sigma_z = 32$ m; $x = 12$ km, $\sigma_z = 154$ m; $x = 100$ km,

$\sigma_z = 460$ m $\sigma_z = 32 x^{0.63}$ for $x = 1\text{-}12$ km

$\sigma_z = 42.7\, x^{0.52}$ for $x = 12\text{-}100$ km

3. $X = (106.95\, Q/u)x^{-1.41} \exp(-0.000274\, H_e^2 x^{-1.04}) = ax^{-1.41}\exp b$

$dX/dx = -1.41\, ax^{-2.41}\exp b + 1.04\, x^{-2.04}(0.000274\, H_e^2 ax^{-1.41}$

$(\exp b) = 0$

$x^{1.04} = 0.000285\, H_e^2/1.41$ $x = 0.00028\, H_e^{1.923}$

Check against Problem 1: $H_e = 400$ m $x_m = 28.2$ km

4. At 160°C, $Q_H = 3.23(10^7)$ cal/sec. At 150°C, $Q_H = 3(10^7)$ cal/sec

   $0.23(10^7)91.8/454)(3.16 \times 10^7)(\$0.75/10^6) = \$216,000/yr$

5. $\Delta H_2 = 429$ m, $v_2 d_2 = 267$ m^2/sec; $\Delta H_1 = 403$ m, $v_1 d_1 = 201$ m^2/sec

   $v = k/d^2$; $d_2 = (201/267)24 = 18.1$ ft $= 5.51$ m

   $v_2 = 267/5.51 = 48.4$ m/sec; $v_1 = 27.4$ m/sec

   Velocity head: $v = (66.7/3.28)\sqrt{h(423/293)}$ $h = (v/24.4)^2$

   $h_2 = (48.4/24.4)^2 = 3.92$ in. w.g.; $h_1 = 1.26$ in. w.g.

   Friction head: $h = 0.0942(°R/D^4)(Q_m/10^5)^2(1 + fH_b/D)$

   $150°C = 762°R$, use $°R = 770$; $Q_m = 7.62(10^6)$ lb/hr

   Assume $H_b = 370$ ft; use $D_2 = 26$ ft and $D_1 = 20$ ft

   $N_{Re\ 2} = 1.37(10^7)$, $f_2 \approx 0.009$; $N_{Re\ 1} = 1.05(10^7)$, $f_1 \approx 0.01$

   $h_2 = 3.07$ in. w.g.; $h_1 = 1.05$ in. w.g.

   Annual cost: HP $= Qh/(6356\,E)$ Assume $E = 0.60$

   HP $= 641\ h$ Use \$100/yr-HP for estimating.

   Cost 2 $= 641(3.92 + 3.07)(\$100) = \$448,000/yr$
   Cost 1 $= 641(1.26 + 1.05)(\$100) = \$148,000/yr$

   Additional cost $= \$300,000/yr$

6. 10°C gave $\Delta H = 26$ m or 18°F for 85.3 ft.   1°F for 4.74 ft of ΔH.

   Depends on the relative sizes of the jetting and buoyancy terms.

7. $H = 400$ ft $= 122$ m $R = 22.2/(3.28 \times 2) = 3.38$ m

   $v = 90/3.28 = 27.4$ m/sec $\rho_a = 1205(528/550) = 1157$ gm/m^3

   $\rho_s = 1205(528/795) = 800$ gm/m^3

   $F = gvR^2(\rho_a - \rho_s)/\rho_a = 9.81(27.4)(3.38^2)(1157 - 800)/1157$

   $= 949$ m^4/sec^3

   At $x = 3000$ m, $\Delta H = 2.50(3000^{0.56})(949^{1/3})/3.8 = 572$ m

8. $Q = 811,000(0.008)(2)(385/35.3 \times 64)/3600 = 0.614$ Nm^3/sec

   $H_e = 694$ m $x_m = 0.00028(694^{1.923}) = 81.5$ km

   $X_m = 0.614(10^6)/[\pi(3400)(415)(3.8)]\exp[-694^2/(2 \times 415^2)]$

   $= 0.009$ ppm$_v$

CHAPTER 16

1. Diameter by flooding velocity: Fig. 5a, flooding velocity $= 0.365$

   $G_f' = 0.365/0.0334 = 10.93$ fps   Use $G' = 60\%\ G_f' = 6.56$ fps

   $D = [4(3730/60)/(6.56\pi)]^{0.5} = 3.47$ ft

Diameter by head loss:

$$G = \left[\frac{0.031(0.0697)(62.4 - 0.07)(32.2)}{45(0.656)^{0.1}}\right]^{0.5} = 0.317 \text{ lb/sec-ft}^2$$

$A = (260/60)/0.317 = 13.67 \text{ ft}^2$ $D = 4.17 \text{ ft}$

In the absence of pilot plant data, assume H_{OG} and N_{OG} are the same for Berl saddles as for Raschig rings.

2. $H_{OG} = G/(K_G a\,P)$ $3.62 = 1070/(K_G a \times 1)$ $K_G a = 296$ lb/hr-ft^3-atm

3. Assume gas is cooled to 125°F (52°C), moisture is raised from 8 to 13%v, and concentration of sulfur dioxide remains at 2400 ppmv before entry to tower.

    Volume to tower: $2(10^6)(585/760)(1.05) = 1,616,000$ acfm

    $1,616,000(2400 - 400)(10^{-6}) = 3230 \text{ ft}^3/\text{min } SO_2$ removed

    Minimum liquid flow: H = 54 atm/mole fraction @ 40°C$_{avg}$.

    $2400(10^{-6}) = 54x$ $x = 44.4(10^{-6})$

    $$\frac{44.4(385)(585/528)(10^{-6})}{18} = 1054 \text{ ft}^3 SO_2/10^6 \text{ lb water}$$

    Obviously, it is not practical to use 3 million pounds of water per minute in a tower.

    Alkaline water could be used under certain circumstances, e.g., the Battersea process in London and some of the processes under development in the United States. The Battersea process uses 58,000 lbs of water per minute for 333,000 acfm of stack gas.

4. Alkaline water will absorb sulfur dioxide much better than neutral water. Calcium sulfite and calcium sulfate are formed. The marble bed has been used on the premise that the marbles would rub against each other during scrubbing and thereby remove the scale from the marbles. This is a process that is receiving wide experimentation. The lime or limestone is often added in the firebox rather than in the scrubbing liquor.

5. Reduces water requirements to consumptive use. Alleviates problem of treating large volumes of scrubbing liquor for disposal within water pollution regulations.

6. Reference: L. Barbouteau and R. Galaud, "Regenerative Processes for Gas Absorption," Processes for Air Pollution Control, 2nd ed. (G. Nonhebel, ed.), Chapter 7, CRC Press, Cleveland, Ohio, 1972.

    Fig. 7-28. Gas is demisted; H_2S/CO_2 absorbed in plate tower with MEA; gas demisted for discharge and MEA liquor sent to regeneration; MEA solution degassed by reducing pressure (hydrocarbons); solution regenerated by heating in plate tower; H_2S/CO_2 to flare or to sulfur removal process; solution returned to storage for reuse.

Reference: B. R. Carney, "Shamrock Completes Reconversion Program, " Oil and Gas J., 46:17, 56-63, August 30, 1947.

Girbotol gas-treating unit design data:
65,000-100,000 Mcf/day; 17% monoethanolamine in water; 5 units to 50 million each; 7 ft diameter by 68 ft high; 23 trays; solution to 4th tray; water wash to top tray; absorber pressure 200 psig; still pressure 12 psig; still temperature 250°F at bottom and 240°F at top; steam consumption not to exceed 3.2 lb/Mcf of gas; inlet H_2S 160-180 gr/100 ft^3; outlet H_2S 0.02-0.3 gr/100 ft^3; solution rate 2-3 gal/Mcf; CO_2 content 0.3-0.4%$_v$.

CHAPTER 17

1. $x/M = kp^{1/n}$ $k = x/M$ at $p = 1$

2. The curve of Freundlich's isotherm rises faster for $n > 1$ than for $n \leq 1$; therefore, the x/M will be greater for the low pressures (concentrations) found in air pollution control. Conversely, the curve rises slowly when $n < 1$ and x/M is low.

3. $1000 \ m^2/gm \ (3.28^2)(454)/43,560 = 112$ acres/lb

4. Assume 100% collection and NTP. C_2Cl_4 mol. wt. = 166
    $T(60)(1)(20/10^6)(166,385) = 0.08(43)(1.5)(1)$ $T = 9970$ min

5. $1/x = [(1 - fp)/(fx_m)] \ 1/p$ No value for $p = 0$. Slope varies (not a straight line); if $fp \ll 1$, slope approximately $1/fx_m$.

6. Brunauer et al., J. Amer. Chem. Soc., 62, 1723, 1940.

    Type I: Monolayer adsorption. Freundlich $n > 2$. O_2 on carbon black @ -183°C.
    Type II: Monolayer followed by indefinite multilayer adsorption. Freundlich does not hold to saturation; there is upturn in curve. H_2O on carbon black @ 30°C.
    Type III: Multilayer adsorption. Freundlich $n < 1$. Br on SiO_2 @ 20°C.
    Type IV: Monolayer followed by definite multilayer. Curve has a plateau, rises and levels off again before saturation. H_2O on activated carbon @ 30°C.
    Type V: Multilayer. Starts $n < 1$, then changes to $n > 1$ and levels off before saturation. H_2O vapor on activated carbon @ 100°C.

7. Perchloroethylene has higher molecular weight (166) than toluene (92), b.p. 121°C versus 110°C, and slower desorption. The higher latent heat of the toluene requires more steam for desorption.

CHAPTER 18

1. $90/[0.24(0.075)] = 5000^{\circ}F$

2. $\sim4300^{\circ}F$. The expansion of the flame front. The heat content may be for the higher heating value when hydrogen is present.

3. Need ~90 Btu/Ncf for steady flame. Assume gas stream has sufficient oxygen.

   Heat required $= 1335(528/610)(90 - 9) = 93,600$ Btu/min

4. Assume 20 Ncfm of natural gas needed, part of gas stream for combustion air, and natural gas is 100% methane.

   $$CH_4 + 2\ O_2 \longrightarrow 2\ H_2O + CO_2 \quad \text{3 vols: 3 vols}$$
   Use $c_p = 0.24$ Btu/lb-$^{\circ}F$ and $\rho = 0.075$ lb/ft^3 (as air).
   $$[20 + 1335(528/610)]0.075(0.24)(700^{\circ} - 150^{\circ}F)$$
   $$- 1335(528/610)(9) = 1240\ \text{Btu/min}$$

5. Yes. Theoretically, <12% heat recovery would suffice.

6. There will be less incineration, except where boilers are used for other purposes and the waste gas stream can be used as combustion air to the boilers. It is highly probable that regulations against fine particles (<0.1 μm) will be set. Combustion produces many fine particles.

7. A heat exchanger with a coolant whose boiling point is less than $-152.9^{\circ}C$ could be used; lox or liquid nitrogen as coolant. A combination of temperature and pressure could be used. The design parameters include: area of contact, pressure, temperature, and time of contact.

CHAPTER 19

1. Assume intermediate zone:
   $$u_t = 0.20(0.9 \times 981)^{2/3}(0.1)/[1.13(10^{-3})1.89(10^{-4})]^{1/3}$$
   $$= 308\ \text{cm/sec} \quad N_{Re} = 180;\ \text{intermediate zone.}$$
   Practical $u_t = 0.5 \times$ theoretical $= 154$ cm/sec $\simeq 5$ ft/sec
   Volume $= H(1.2H)(10H) = 12\ H^3$
   Residence time $= 12\ H^3/(800/60) = H/5$
   $H^2 = 0.22 \quad H = 0.47$ ft $\quad W = 0.57$ ft $\quad L = 4.7$ ft
   Probably design to use turbulent settling and d_{99}:
   $$q = 0.01 = \exp(-308t/H) \quad t/H = 0.015$$
   $$H^4 = 74.1 \quad H = 2.93\ \text{ft} \quad W = 3.52\ \text{ft} \quad L = 29.3\ \text{ft}$$

2. Collection efficiency for particle size smaller than 1000 μm should be ratio of settling velocities; i.e., for 200 μm,

$$\eta = 200/1000 = 0.20 \text{ or } 20\% \qquad N_{Re} = 7.16; \text{ intermediate zone.}$$

3. s.f. $= v^2/gR = 70^2/(32.2 \times 1.5) = 101$

In the intermediate regime, velocity is proportional to the 2/3rds power of force.

$$\omega = 101^{2/3} u_t = 21.7 \left\{ \frac{0.2(0.9 \times 981)(0.02)}{[(1.13 \times 10^{-3})(1.89 \times 10^{-4})]^{1/3}} \right\}$$

$$= 1335 \text{ cm/sec} \qquad N_{Re} = 156; \text{ intermediate zone.}$$

4. Inlet: $800/(b \times 2b) = 70 \times 60$ \qquad $b = 0.309$ ft

$$d_{100} = \left\{ \frac{18(1.89 \times 10^{-4}) \ln[1.5/(1.5 - 0.309)]}{0.9(46.7^2)(0.7)} \right\}^{0.5} = 7.56(10^{-4}) \text{cm}$$

Angular velocity $(\omega_a) = 70/1.5 = 46.7$ radians/sec

For turbulent flow, effective migration time at 99% collection
$= 0.7 \ln 100 = 0.152$ sec \qquad $d_{99} = 16.4$ μm

5. $d_{100} = [9(1.89 \times 10^{-4})(0.309)/(\pi \times 6 \times 70 \times 0.9)]^{0.5}$

$= 6.65(10^{-4})$ cm $= 6.65$ μm

For turbulent flow, effective b for 99% collection $= 0.309$
$\ln 100 = 1.42$ ft \qquad $d_{99} = 14.4$ μm

6. Laminar regime: $d_{50} = 7.56/\sqrt{2} = 5.35$ μm

Turbulent regime: $d_{50} = 16.2/\sqrt{2} = 11.4$ μm

CHAPTER 20

1. Particles from smoking meat are greasy and sticky.
Particles from slaking lime are wet and sticky (hygroscopic).
Fly ash from coal burning is abrasive, sometimes on fire.
Penetration of radioactive fumes would be greater than the allowable levels.
A heavy mist from a process would wet filter and "blind" it to the passage of air.

2. Area $= \pi (18/36)(30/3) = 15.7$ yd^2

Wt/area $= 18/16 + (1/16 \times 1/12 \times 9)(1 - 0.4)(9.3 \times 62.4)$

$= 17.45$ lb/yd^2

Total weight $= 15.7 (17.45) = 274$ lb/bag

3. Fabric cost $= \$11.50$/yd^2 \qquad (from Table 20-1)

Cost per bag: $\$11.50(24/9) + \$5.50(1.25)(24/20)$

(from Sec. V, Costs + inflation) $= \$38.92$

Labor @ $3.75/hr: 20 min for $1.25

Replacement cost: 2000($38.92 + $1.25) = $80,340

4. Plenum area = 400,000/(5 X 20) = 4000 ft^2

Baghouse volume = 4000(18) = 72,000 ft^3

Electrostatic precipitator volume:

$q = 0.005 = \exp[-0.33t/(4.8/12)]$ $t = 6.42$ sec

Volume = 400,000(6.42/60) = 42,800 ft^3

5. $W \simeq 2/3 L$ $2/3 L^2 = 4000$ $L = 77.5$ ft for baghouse

$WH = 400,000/(60 \times 5) = 1330$ $H = 20$ ft $W = 67$ ft

$L = 6.42(5) + 1$ ft for charging = 33 ft

Retention times:
Baghouse = 72,000/(400,000/60) = 10.8 sec
Electrostatic precipitator = 6.42 + 0.2 = 6.6 sec

6. Use annual capital cost = 15% of first cost.

$$\left[\frac{50,000\alpha/v}{50,000\alpha/5}\right]^{0.9} = \frac{C}{2.5(50,000)} \qquad C_c = \frac{79,900}{v^{0.9}}$$

Pumpage costs:

$C_p = 0.746(0.015)(50,000)(h)(8760)/(6356 \times 0.60) = 1285\,h$

$h = 1.2$ in. w.g./fpm $C_p = 1542\,v$

Minimum cost:

Cost = $C_c + C_p$ (Assume that other costs are equal.)

$dC/dv = 1542 - 71,900/v^{1.9} = 0$ $v = 7.55$ ft/min

CHAPTER 21

1. Approximately $q = 1.59\exp(-1.59\,d)$

2. Graph.

3. The smaller the particles, the lower in the plot (lower N_t) the line will occur and the more uniform the particle size, the flatter the slope of the line will be.

4. $q = \exp(-61.2\,v_g tL'/D)$

Equation 21-6 shows penetration (efficiency) to be independent of particle size; such is not the real situation. Particle diameter should appear in the denominator of the exponential term; i.e., the impaction parameter or some similar expression should be in the exponential.

5. @ 20°C
$$D = \frac{585}{400/3.28}\sqrt{72.8/1} + 597[0.01005/\sqrt{78.1(1)}]^{0.45}$$
$$[1000(6)/(7.48 \times 1000)]^{1.5} = 61.6\,\mu m$$

@ 80°C
$$D = \frac{585}{400/3.28}\sqrt{62.6/1} + 597[0.00357/\sqrt{62.6(1)}]^{0.45}$$
$$[1000(6)/(7.48 \times 1000)]^{1.5} = 51.3\,\mu m$$

6. $q = \exp[-0.125(245)(6)/51.3] = 0.028$
$$\psi = \frac{1.11(2)(400 \times 30.48)(10^{-4})2}{18(2.11 \times 10^{-4})51.3(10^{-4})} = 13.89$$

$q = \exp[-0.09(6)\sqrt{13.89}] = 0.13$

$d_{50} = \exp\{-1.56 - 1.46\ln[(400/66.7)^2/5]\} = 0.012\,\mu m$

$$q = \exp\left\{\frac{2(51.3 \times 10^{-4})(1)(400 \times 30.48)}{55(2.11 \times 10^{-4})}\frac{6/7.48}{1000}\frac{1}{13.89}\right.$$
$$\left.\left[-0.7 - 13.89(0.45) + 1.4\ln\left(\frac{13.89 + 0.7}{0.7}\right) + \frac{0.49}{13.89(0.45) + 0.7}\right]\right\}$$

$= \exp(-1.636) = 0.195$

$P = 1000(36)/(6356 \times 0.60) = 9.44$ HP/1000 Ncf $= 7.84$ HP/1000 acf

$N_t = 1.25(7.84)^{0.6} = 4.30$

$q = \exp(-4.30) = 0.014$

7. The equations were empirically derived, usually on heterogeneous
    dusts, and their application requires a great deal of judgment
    and experience for any degree of success.

CHAPTER 22

1. $u_t = 0.003\rho_p d^2 = 0.003(2.7)(1)^2 = 0.0081$ cm/sec
    Gravities $= 6.91/0.0081 = 853$ $N_{Re} = 0.0023$; laminar regime.

2. Graph.

3. Graph. Yes. The drag coefficient changes with the Reynolds
    number and, therefore, with velocity.

4. Graph.

5. Installed cost: $\$3.25(80,000)(2.25) = \$585,000$
    Capital costs: $\dfrac{585,000}{15}\dfrac{(1 + 0.08)^{15}}{(1 + 0.08)^{15} - 1} = \$56,950$
    Power costs:

    Corona power $= 390$ W/1000 acfm (80,000 acfm) $= 31.2$ kW

    Head loss $= 0.746(80,000)(1.6)/(6356 \times 0.60) = 25.04$ kW

Total power costs: $(31.2 + 25.04)(8760)(0.016) = \$7880/\text{yr}$

Maintenance costs: $0.032(80,000) = \$2560/\text{yr}$

Total annual costs: $56,950 + 7880 + 2560 \approx \$67,400$

6. Graph.

7. $4(10^6)(8)/60 = 533,000 \text{ ft}^3$

$6\,WH = 4(10^6)/60 \qquad WH = 11,100 \text{ ft}^2$

$H = 60$ ft (3 20 ft heights + hopper) $\quad W = 184$ ft

$L = 48$ ft + 1 ft charging = 49 ft $\quad$ Use 50 ft

Sections: Width--8 X 23 ft
$\qquad\qquad$ Length--3 X 16.7 ft

CHAPTER 23

1. Total value of material:

$10,000(100/7000)(1440 \text{ X } 365)(80/2000) = \$3.00(10^6)/\text{yr}$

Cost of cyclone:

$\$11,280$ (from Table IV plus 25% inflation)

Value of recovered material:

1st cyclone: $0.84(3.00 \text{ X } 10^6) = \$2,520,000/\text{yr}$

2nd cyclone: $0.3(0.84)(1.00 - 0.84) = 0.04$

$\qquad\qquad\qquad\qquad\qquad = \$120,000/\text{yr}$

3rd cyclone: $0.3(0.3 \text{ X } 0.84)(1.00 - 0.84 - 0.04) = 0.0091$

$\qquad\qquad\qquad\qquad\qquad = \$27,300/\text{yr}$

4th cyclone: $0.3(0.3^2 \text{ X } 0.84)(1.00 - 0.84 - 0.049) = 0.0025$

$\qquad\qquad\qquad\qquad\qquad = \$7,500/\text{yr}$

2. The principal advantage gained by putting a multiple cyclone in series with an electrostatic precipitator is that some particles with excessive bulk resistivities may be large enough for collection by the multiple cyclone. The particles of fly ash will have good bulk resistivities for collection by the electrostatic precipitator on the hot side; however, the volume of gas to be treated will be increased by nearly 60%, and so will the size of the electrostatic precipitator.

3. Reference: W. Strauss, Industrial Gas Cleaning, Pergamon, London, 1966, p. 223.

$x = (4\not{D}t/\pi)^{0.5} \quad \not{D} = 6.82(10^{-6}) \text{ cm}^2/\text{sec}$ (Table 5-2)

x = distance cleared in time t

Use laminar flow regime and 0.25 in. spacing of ribbons.

$$N_{Re} = 3 = 0.0012(2.54/4)v/(1.8 \times 10^{-4}) \quad v = 0.71 \text{ cm/sec}$$

$$\frac{10,000/60}{WH} = \frac{0.71}{30.48} \quad WH = 7168 \text{ ft}^2$$

$$2.54/4 = (4 \times 6.82 \times 10^{-6}t/\pi)^{0.5} \quad t = 46,400 \text{ sec}$$

Chamber dimensions: 20 ft high X 360 ft wide X 1080 ft long

4. The advantages would be high efficiency of removal on even
   small particles with low head loss. The disadvantages are
   the size of such a unit and the difficulty of removal of
   particles from the ribbons. The size calculated above may
   be compromised some by accepting the intermediate flow
   regime.

CHAPTER 24

1. D. H. McCrea, J. H. Field, and E. R. Bauer, Jr., "The Alka-
   lized Alumina System for SO$_2$ Removal: Design and Operation
   of a Continuous Pilot Plant," ASME Preprint 68-WA/FU3,
   Amer. Soc. of Mech. Engr., New York, 1968.
   Control Techniques for Sulfur Oxide Air Pollutants, National
   Air Pollution Control Administration Publ. No. AP-52, 1969.

   1/16-in. spheres of dawsonite [NaAl(CO$_3$)(OH$_2$)] activated
   @ 1200°F to give high-porosity, high-surface area sodium
   aluminate; solids react with SO$_2$ @ 300°-650°F to form
   sodium sulfate; solids regenerated with reducing gas @ 1200°F;
   (producer gas used for reducing--H$_2$, CO, and CO$_2$); solids
   returned to process; H$_2$S sent to Claus process for conversion
   to elemental sulfur.

   Process fell into disfavor because of sorbent losses.

2. NO = 291 + 3.67(400/2) = 1025 ppm$_v$

   NO = (°F/5340)$^{0.535}$(% XS air + 38)

   = (2800/5340)$^{0.535}$(30 + 38) = 48.1 lbs/10^6 Btu fired

   $$\frac{48.1(385/30)}{10^6(12,500 \text{ Nft}^3/10^6 \text{ Btu})} \frac{10^6}{\text{ft}^3/\text{ft}^3} \text{ppm}_v = 1027 \text{ ppm}_v \quad (\text{see Ex. 2-2})$$

   Very close agreement is quite accidental; neither formula is
   anywhere near that exact.

3. In the first step, the weight was reduced by burning more HC
   and CO through increasing air-to-fuel, also increased NO$_x$.

   The next step requires a thermal converter that will be rather
   costly to install and maintain.

   The final step will need catalytic converters after burning the
   HC and CO to break down the NO formed.

The equipment required gets more expensive with each step. This follows the general rule that the last vestige of any pollutant is difficult and expensive to remove.

4. Assume that sulfuric acid mist approximately same as phosphoric acid mist.

$$N_t = 1.33 P^{0.647} \qquad \text{(Eq. 21-1 and Table 21-1)}$$

$$N_t = 1.33(7.5/10)^{0.647} = 1.10$$

$$q = \exp(-N_t) = \exp(-1.10) = 0.33 \qquad \eta = 67\%$$

5. Fan blades designed by manufacturer for quieter operation than has been the practice.

   Fan set on vibration-damping mounts (lead, rubber, plastic, or springs).

   Ductwork connected to fan with flexible materials.

   Ductwork near fan, especially discharge side, made of sound-absorbing duct material. Duct has acoustical insulation sandwiched between outside of duct and perforated inner liner.

Appendix C

SOME SOURCES OF AIR POLLUTION EQUIPMENT

I. SAMPLING EQUIPMENT

A Pumps
B Meters
C Paper tape samplers
D Filters
E Impingers

F Impactors
G Electrostatic precipitators
H Thermal precipitators
I Stack sampling trains
J Meteorological

Andersen 2000 Inc.   FI
P. O. Box 20769
Atlanta, Georgia 30320
[(404) 763 2671]

Gelman Inst. Co.   ABCDE
600 S. Wagner Rd.
Ann Arbor, Mich. 48106
[(313) 665 0651]

Bendix: Env. Sci. Div. ABCEGIJ
1400 Taylor Avenue
Baltimore, Maryland
[(301) 825 5200]

Mine Safety Appl. Co. ADEGH
400 Penn Center Blvd.
Pittsburgh, Pennsylvania 15235
[(412) 241 5900]

Curtin Scientific Co.   ABCDE
P. O. Box 1546
Houston, Texas 77001
[(713) 923 1661]

Research Appl. Co. ABCDEFI
Allison Park,
     Pennsylvania 15101
[(412) 961 0588]

II. ANALYTICAL EQUIPMENT

A Electrical sulfur dioxide
B Chemiluminescence
C Gas chromatographs
D UV-Vis. spectrophotometers
E Infrared spectrophotometers
F Noise

G Mass spectrometers
H Transmissometers
I x Ray fluorescence
J Calibration
K Particle size/mass
L Atomic absorption

Analytical Inst. Devel.   BCJ
250 S. Franklin Street
West Chester, Pa. 19380
[(215) 692 4575]

GCA Technology Div.   K
Burlington Rd.
Bedford, Mass. 01730
[(617) 275 9000]

Bailey Meter Co.   H
29801 Euclid Avenue
Wickliffe, Ohio 44092

General Radio Co.   F
300 Baker Avenue
Concord, Mass. 07142
[(617) 369 8770]

B & K Instruments, Inc.   F   Micromeritics Inst. Corp.   K
5111 West 164th Street   800 Goshen Springs Rd.
Cleveland, Ohio 44142   Norcross, Georgia 30071
[(216) 267 4800]   [(404) 448 8282]

Barringer Research Ltd.   DE   Perkin-Elmer Corp.   BCDEGL
304 Carlingview Drive   Main Avenue
Rexdale, Ontario, Canada   Norwalk, Conn. 06856
[(416) 677 2491]   [(203) 762 1000]

Bausch and Lomb   K   Philips Electronic Inst. ACDE
820 Linden Avenue   750 S. Fulton Avenue
Rochester, New York 14625   Mt. Vernon, New York 10550
[(716) 385 1000]   [(914) 664 4500]

Beckman Inst., Inc.   ADEL   REM Scientific, Inc.   B
2500 Harbor Blvd.   2000 Colorado Avenue
Fullerton, California 92634   Santa Monica, Calif. 90404
[(714) 871 4848]   [(213) 828 6078]

Cahn Instruments   K   Royco Instruments, Inc.   K
7500 Jefferson Street   141 Jefferson Drive
Paramount, California 90723   Menlo Park, Calif. 94025
[(213) 634 7840]   [(415) 325 7811]

Columbia Scientific Inst.   I   Thermo-Systems, Inc.   K
3625 Ed Bluestein Blvd.   2500 Cleveland Ave. No.
Austin, Texas 78762   St. Paul, Minnesota 55113
[(512) 926 1530]   [(612) 633 0550]

duPont Inst. Prod. Div.   DE   Tracor, Inc.   CJ
1007 Market Street   6500 Tracor Lane
Wilmington, Delaware 19898   Austin, Texas 78721
[(302) 453 2711]   [(512) 926 2800]

Bendix (see above)   BCH   MSA (see above)   DE

III. GASEOUS CONTROL

A Afterburners   D Adsorbers
B Catalytic units   E Odor controls
C Absorbers   F Fans and blowers

Air Correction Div. UOP   A   Green Fuel Econ. Co.   F
Tokeneke Road   627 Main Street
Darien, Connecticutt 06820   Beacon, New York 12508
[(203) 655 8711]   [(914) 831 3600]

Airkem	E	Oxy-Catalyst, Inc.	AB
111 Commerce Road		E. Biddle Street	
Carlstadt, New Jersey 07072		West Chester, Pa. 19380	
[(201) 933 8200]		[(215) 692 3500]	

American Air Filter	D	Pitt ACD: Calgon Corp.	D
215 Central Avenue		P. O. Box 1346	
Louisville, Ky. 40208		Pittsburgh, Pa. 15230	
[(502) 637 0011]		[(412) 923 2345]	

Buffalo Forge Co.	F	Research-Cottrell, Inc.	A
490 Broadway		P. O. Box 750	
Buffalo, New York 14240		Bound Brook, N. J. 08805	
[(716) 847 5121]		[(201) 885 7000]	

Carus Chem. Co.	E	The W. W. Sly Mfg. Co.	C
1500 Eighth Street		4700 Train Avenue	
LaSalle, Illinois 61301		Cleveland, Ohio 44101	
[(815) 223 1500]		[(216) 631 1100]	

Catalytic Prod. Int'l	B	J. P. Stephens & Co.	C
524 Mill Valley Road		1185 Ave. of the Americas	
Palatine, Illinois 60067		New York, New York 10036	
[(312) 359 0403]		[(212) 575 2000]	

Engelhard Industries	B	Union Carbide: Carb. Prod.	D
430 Mountain Avenue		270 Park Avenue	
Murray Hill, N. J. 07974		New York, New York 10017	
[(201) 242 2700]		[(212) 551 2345]	

IV. PARTICULATE CONTROL

A Inertial cleaners   E Fiber filters
B Centrifugal cleaners   F Scrubbers
C Fabric filters   G Electrostatic precipitators
D Baghouses   H Sonic precipitators

American Air Filter	ABCDEFG	Research-Cottrell	BCDEFG
(see above)		(see above)	

Bemis Co.	C	Koppers Co.	G
P. O. Box 3758		200 Scott Street	
St. Louis, Missouri 63122		Baltimore, Md. 21203	
		[(301) 727 2500]	

Buell/Envirotech	ABG	Mahon Indust. Corp.	BCDFG
253 N. Fourth Street		6045 Dixie Hwy.	
Lebanon, Pa. 17042		Saginaw, Mich. 48606	
[(717) 272 2001]		[(517) 777 2521]	

Carter-Day Co.   BCDE
655 19th Avenue, NE
Minneapolis, Minn. 55418
[(612) 781 6541]

The Ducon Co.   BCDF
147 E. Second Street
Mineola, New York 11501
[(516) 741 6100]

DuPont   ACE
Centre Road Bldg.
Wilmington, Del. 19810
[(302) 999 3135]

Dustex Corp.   ABCDFG
2777 Walden Avenue
Buffalo, New York 14225
[(716) 685 1200]

Globe Albany Corp.   C
1400 Clinton Street
Buffalo, New York 14240

Mikropul/U. S. Filter ABCDEFG
8 Chatham Road
Summit, New Jersey 07901
[(201) 273 6360]

Peabody Engr. Corp. ABCDEFG
835 Hope Street
Stamford, Conn. 06907
[(203) 327 7000]

The Torit Corp.   BCD
1133 Rankin Street
St. Paul, Minnesota 55116
[(612) 698 0391]

Western Precip.: Joy ABCDFG
P. O. Box 2744 Term Annex
Los Angeles, Calif. 90051
[(213) 627 4771]

Wheelabrator-Frye, Inc.   CD
930 Ft. Duquesne Blvd.
Pittsburgh, Pa. 15222
[(704) 391 8960]

AUTHOR INDEX

Underlined numbers refer to Bibliography pages, where complete citations are given.

A

Adams, R. E.	_133_
Amdur, M. O.	8
Anderson, W. L.	168

B

Barnard, R. E.	_249_
Barneby, H. L.	_133_
Bartok, W.	238, _248_
Baxter, W. A.	_220_
Beranek, L. L.	_248_
Billings, C. E.	183, 186, _187_
Borgwardt, R. H.	179, _187_
Brandt, A. D.	53, _69_
Brief, R. S.	_187_
Briggs, G. A.	_95_
Browning, W. E., Jr.	_133_
Brunauer, S.	124, _133_
Busby, H. G.	_221_

C

Calvert, S.	_120_, 195, _202_
	_203_
Carlton-Jones, D.	91, _95_
Cassell, E. J.	_20_
Carpenter, S. B.	_95_, _96_
Chapman, R. L.	_20_

Citarella, J. F.	_96_, _221_
Chilton, C. H.	_96_, _121_, _149_
Chopey, N. P.	248
Colbaugh, W. C.	_95_
Constance, J. D.	_69_
Cottrell, F. G.	_205_
Crawford, A. R.	_248_
Crawford, W. D.	_221_
Crowell, A. D.	_133_

D

Danielson, J. A.	_70_, _121_, _133_
	_149_, _166_, 208, _221_
Darby, K.	_221_
Davies, C. N.	_188_
Davis, J. C.	_249_
Dennis, C. S.	_249_
Deutsch, W.	209
Dorman, R. G.	_188_
Drinker, P.	_166_
Durham, J. F.	179, _187_

E

Eckert, J. S.	108
Emmett, P.	124, _133_
Ergun, S.	128

F

Fair, J. R.   107, 121
Field, R. S.   203
First, M. W.   74, 78, 161
                            166
Freundlich, H.   124
Friedlander, S. K.   166, 169

G

Gartrell, F. E.   96
Gibbs, J. W.   124
Gieseke, J. A.   196
Gifford, F. A.   81, 95
Greco, J.   221

H

Harrington, R. E.   188
Harris, L. S.   203
Hayden, P. L.   249
Hemeon, W. C. L.   70
Herrick, R. A.   188
Hirai, M.   249
Holland, J. Z.   87
Horlacher, W. R.   249

I

Iinoya, K.   188
Ingraham, T. R.   249

J

Johnstone, H. F.   194, 195
                            203
Johnston, W. A.   133
Junge, C. E.   228

K

Katz, J.   221

Kirkpatrick, S. D.   96, 121
                            149
Kopita, R.   203
Küster, R. W.   124

L

Lancaster, B. W.   203
Langmuir, I.   124
Lapple, C. E.   158, 161, 166
Larsen, R. I.   20
Ledbetter, J. O.   21, 249
Leva, M.   121
Liddle, C. J.   121
Löffler, F.   188
Lund, H. F.   47
Lundgren, D. A.   184, 203

M

Mackinnon, D. J.   249
Manny, E. H.   248
McCaldin, R. O.   96
McLaughlin, J. F.   221
Mehta, D. S.   203
Montgomery, T. L.   88, 95

N

Nakada, Y.   215
Nelson, F.   95
Nonhebel, G.   47, 163
Nukiyama, S.   193, 203

O

O'Connor, J. R.   96, 221
Odello, R.   249
Oglesby, S., Jr.   221
Orr, C., Jr.   188

P

Pack, D. H.   95
Paretsky, L.   188
Pasquill, F.   81, 96
Patty, F. A.   21
Penney, G. W.   221
Perry, J. H. 90, 91, 96, 121
                     149, 158
Perry, R. H.   96, 121, 149
Pfeffer, R.   188
Phillips, P. E.   188
Piegari, G. J.   248

R

Ramskill, E. A.   168
Randall, C. W.   249
Ranz, W. E.   121
Reese, J. T.   221
Rich, L. G.   121
Roberts, M. H.   194, 195
                          203
Robinson, J. W.   188
Robinson, M.   221
Rose, A. H.   76, 78, 187
Ross, R. D.   47, 133, 149
                          249
Ruff, R. J.   140

S

Scheele, C. W.   124
Schneider, H. B.   91, 95
Schrader, K.   221
Semrau, K. T. 190, 191, 203
Shah, I. S.   249
Shenfeld, L.   95

Sherwood, T. K.   108, 121
Shimamura, H.   249
Silverman, L.   166
Smith, S. E.   188
Spaite, P. W.   181, 188
Sporn, P.   96
Sproull, W. T.   215
Squires, A. M.   188
Stairmand, C. J.   153, 154
          163, 166, 190, 203
St. Clair, H. W.   228
Stenburg, R. L.   78
Stephan, D. G. 78, 187, 188
Stephens, N. T.   96
Stern, A. C. 47, 70, 140, 188
                          228
Strauss, W.   121, 133, 149
      161, 166, 188, 203, 221
                     228, 249

T

Tanasawa, Y.   193, 203
Tassler, M. C.   203
Teague, R. K.   249
Teller, A. J.   121, 124
Theodore, L.   188
Thomas, F. W.   95, 96
Treybal, R. E.   121
Turner, D. B. 82, 84, 85, 86
                          96

V

Veldhuizen, H.   249

W

Walsh, G. W.   181, 188
Whitby, K. T.   184

White, H. J.   215, <u>221</u>   Y
White, P. A. F.   <u>188</u>   Young, D. M.   <u>133</u>
Wilder, J.   183, 186, <u>187</u>   Z
                                    Zenz, F. A.   <u>121</u>

SUBJECT INDEX

A

Abatement   1, 23, 73
Abrasiveness   66
Absolute filters   173
Absorbents   118
Absorption
  cocurrent   104
  countercurrent   104
  equilibrium   102, 104
  factor   105
  tower size   105-116
Acid production   235
Activated carbon   126
Adsorption
  energy   123
  isotherms   124
  kinetics   125
Adsorptivity   27, 126
Agriculture   74
Ammonia scrubbing   235-236
Arndt-Shulz law   5
Atmospheric stability   82

B

Baffle cleaner   164
Bag filter   183-185

C

Capture velocity   50, 52
Carbon monoxide   6, 239-241
Carboxyhemoglobin   7

Catalytic incineration   138
                         143-146
Chemisorption   123
Clausius-Clapeyron
  equation   125
Coal gasification   232
Collecting electrodes   211-213
Collecting hoppers   66
Combustibility   27
Combustion
  modifications   237-238
  theory   135
Condensation   147, 225
Contacting power   190-191
Control cost   11, 152-154
Cooling hot gases   60
Cooling tower fog   245-246
Corona
  discharge   205
  power   213-214
Correlations   6
Corrosion   27
Costs
  annual   33, 152-154
  comparative   152-154
  estimates   29, 35
  installed   32
  maintenance   34
  operating   30, 33
  versus size   31
Cyclones   159-164

Cyclonic scrubbers   193   Feedlots   246-247

                D   Fiber filters   172, 174, 176

Deflecting vanes   66   Filter
                                           cleaning   176-177
Desulfurization   232   efficiency   177-178
                                           velocity   182
Deutsch equation   209
                                         Fire   27
Dew point   27
                                         Flame incineration   136
Diffusional impaction   169   138-141, 146

Dilution cooling   60   Flare   139-141

Dispersion   36   Flooding velocity   107
   calculations   82, 86
   parameters   83-85   Flow rate   25, 27

Duct design   57   Freundlich isotherm   124

Dynamic projection   161, 164   Fuel substitution   36, 75

                E   G

Effects   5-9   Gas reheating   63

Efficiency   39   Gas transfer unit   105

Ejector venturi   197, 199   Granular filters   173, 176

Electrical migration   206   Gravity scrubbers   189-190

Electrode   Guarantees   43
   cleaning   214-217
   spacing   210   H
   types   211-213
                                         Hazards   27
Emission limits   3, 17
                                         Head loss
Epidemiology   4   absorbers   107, 117
                                           adsorbers   128-129
Equilibrium curve   102, 104   cyclones   161
                                           electrical precipitators
Exhaust   217, 219
   hoods   49   exhaust   53-55
   volume   50-52   filters   178-180
                                           stacks   89
Explosion   27, 66   tray towers   117
                                           venturi scrubbers   196-197
Explosive limits   64
                                         Heat dissipation   129
                F
                                         Henry's law   100
Fabric filters   173-174
                                         HEPA filters   173, 176
Fan
   characteristics   58, 59   Hood entry loss   52-54
   laws   59

Hopper valves   66-67
Horsepower   34
Hot water scrubbing   199
Hydrogen sulfide   118, 232

I

Impaction   167
Impingement   199
Incineration   39ff.
Interception   167-169

L

Land use   3
Langmuir isotherm   124
Lead   239, 242
Lime/limestone scrubbing
                       233-235
Liquid transfer unit   105
Louver separator   165

M

Marble bed scrubbers   200
Materials
   disposal   65-66
   handling   65
   substitution   76
   transport   65
$\mu g/m^3$ 26
Mill tailings   77
Moisture   27
Molecular diffusivity   29
Mole fraction   103
Molten metals   242

N

Ncfm (scfm)   25
Noise control   244

Nuclear energy   74

O

Odors   29, 147-148
Operating lines   104
Oxidant   8
Oxidation   147
Ozone   9

P

Papermaking   74
Particle size   28
Pasquill's equation   81
PCV   239
Penetration
   electrical precipitators
                       209-210
   filters   178, 180
   graphical solution   42
   overall   40
   scrubbers   192-199
   versus particle size 40-43
Physical adsorption   123, 130
Physiological effects   6
Planning   1
Plant damage   8
Plume rise   87-88
Pollen   244-245
Pollutant lifetimes   37, 82
Pollution potential   80
Power generation   77
ppm_v 26
Precleaning   64
Process change   36, 73

Q

Quality factor   180

Quality guides   14   Stagnation   3

Quenching   60   Standards
                                              emergency   15
          R   primary   14
Radiation/convection   61, 62   secondary   14

Radioactive gases   147   Stickiness   66

Regeneration   130   Sulfur dioxide   5, 7, 38
                                                                  231-236
Retentivity   127
                                         Sulfur trioxide   8
Resistivity   214

Ringelmann number   37   T
                                         Tallest stacks   93

                                         Temperature   25
          S
                                         Thermal incineration   138
scfm (Ncfm)   25   141-143, 146

Settling chamber   156-159   Threshold   4

Smelting   74, 242-243   TLV   15

Smokeless flare   139-141   Toxicology   4

Solubility   27, 101   Transfer unit   43

Solvent recovery   127   Transport velocity   55

Sonic agglomeration   224-225   Treatability   27

Specifications   43   V

Spray scrubbers   191-193   Venturi scrubbers   193-197

Stack   Viscous filters   173
    draft   89
    foundation   92   W
    friction   89
    structure   89   Wastewater treatment   246